A REALIST PHILOSOPHY OF ECONOMICS

Karl Mittermaier

Foreword by
Isabella Mittermaier

With an extended preface by
Alan Kirman

and a prologue by
Rod O'Donnell

BRISTOL
UNIVERSITY
PRESS

First published in Great Britain in 2023 by

Bristol University Press
University of Bristol
1–9 Old Park Hill
Bristol
BS2 8BB
UK
t: +44 (0)117 374 6645
e: bup-info@bristol.ac.uk

Details of international sales and distribution partners are available at bristoluniversitypress.co.uk

British Library Cataloguing in Publication Data
A catalogue record for this book is available from the British Library

ISBN 978-1-5292-3440-4 hardcover
ISBN 978-1-5292-3442-8 ePub
ISBN 978-1-5292-3443-5 OA PDF

The right of Karl Mittermaier to be identified as author of this work has been asserted by him in
accordance with the Copyright, Designs and Patents Act 1988.

Cover design: Hayes Design and Advertising
Front cover image: Freepik/Geometric head/economics
Bristol University Press uses environmentally responsible print partners.
Printed and bound in Great Britain by CPI Group (UK) Ltd, Croydon,
CR0 4YY

For Isabella

Contents

About the Author

Karl Hermann Martin Mittermaier (1938–2016) immigrated to South Africa from Berlin, East Germany in 1949. He was a descendant of the great German jurist Karl Joseph Anton Mittermaier (1787–1867). Karl obtained his BCom (1959), BA (1961) and BA(Honours) (1963) all from the University of Cape Town and his PhD (1988) from the University of the Witwatersrand, Johannesburg.

Acknowledgements

I would like to thank the two anonymous referees for their thoughtful comments on Karl's book, their recommendations and their endorsements.

I am extremely grateful to Alan Kirman and Rod O'Donnell for their contributions to this book.

I am also extremely grateful to Eugene le Roux and Michael Stettler whose input has got this book to publication.

Isabella Mittermaier
May 2023

Foreword

Isabella Mittermaier

Karl's interest in the rationale of an empirical method in economics arose when he was employed by a financial services company in the early to late 1960s to prepare economic reports. He realized that, apart from an understanding of certain institutional arrangements such as banking, the economics acquired in a formal education seemed to be of little relevance to the statistical investigations that are done. That was the starting point for this dissertation; the totally different character of theoretical and empirical economic studies, and the fact that very little use seems to be made of economic theory in most empirical studies. There was a question to be investigated; what is the empirical content of economics?

This book is part of that dissertation written in the mid 1970s. Days and nights ran into each other. Karl grew a beard during this period and was never clean-shaven again. I typed, on a portable typewriter, what Karl wrote in pencil on a foolscap pad.

When it was finished Karl asked me to accompany him when he went to hand it to a professor of economics at his home in Johannesburg. This puzzled me but I did go with him and he did hand it to the professor. Periodically I would ask Karl what was happening about it, and he always replied that one person did not understand it at all and the other person only understood 10 per cent of it.

One morning, roughly ten years ago, I got up early and Karl, who had not gone to bed yet, was sitting on the staircase reading this dissertation and he said to me "This is brilliant." Karl did not blow his own trumpet.

Later, when I asked the professor to whom Karl had handed the dissertation what he had done with it, he told me that it was not his field and that he had given it to another professor of economics.

I was never allowed into Karl's study. When I had to sort it out I could not find the dissertation. I had sleepless nights thinking that I had failed Karl. Then I found the chapters. Michael Stettler, a former student and colleague of Karl's, and one of Karl's favourite young men, sorted out the chapters and Karl's other favourite young man, Eugene le Roux, an accountant, spent his entire Christmas break putting it into digital format.

Karl's face used to light up whenever Eugene or Michael came to visit him in his latter years.

Johannesburg
May 2023

Extended Preface: A Realistic Attitude to the Economy

Alan Kirman

I have one regret in writing this little chapter to accompany Karl Mittermaier's book, which is that I never met him. There are many things on which we clearly agreed and many others that I would like to have discussed with him. What is most interesting is that he tackles the fundamental problems of economic theory, the role of time, the nature of equilibrium, what sort of theory could be developed to deal more satisfactorily with these head on. He links these problems with the contributions of a host of predecessors. But, of course, in attempting to do this he did not himself claim to have produced an all-encompassing theory but rather point out what elements such a theory would need.

1. Philosophical considerations

His philosophical bent shows through on practically every page of this book, and he was fascinated by debates in that discipline.

For instance, he spent considerable time on the notion of free will and the debate as to its importance in economics. He characterized two approaches as to how free will enters into the economic picture. The usual discussion as to the nature and role of free will in economics boils down, in his view, to a debate between two different positions. On the one hand there is the 'complexity' approach, which argues that the whole system is 'too complex' for the human mind to grasp and that therefore the future remains essentially unpredictable. The alternative position is associated with the free-will approach. This could be reduced, as Mittermaier explains, to arguing that the only obstacle to explaining the whole system and its functioning is that human action intervenes. This is too simplistic a summary of his view, but his reflection on the subject was enough to lead him to reject the pure free-will argument.

Another debate which particularly interested him was that about nominalism.

Michael Stettler (2019) says that one of Mittermaier's favourite quotes was that where Pareto called himself 'the most nominalist of nominalists'. For Pareto, all good science has to follow the nominalist path, purging itself of any metaphysics. And so, Mittermaier chose as the epigraph for one of his unpublished manuscripts the following from Pareto's (1935) *Mind and Society*:

> Literary economists ... are to this day dilly-dallying with speculations such as 'What is value?' They cannot get it into their heads that things are everything and words are nothing, and that they may apply the terms 'value' and 'capital' to any blessed things they please, if only they be kind enough – they never are – to tell one precisely what those things are. (Pareto, 1935, p 62)

To a non-philosopher, this recalls a famous quotation from *Through the Looking Glass* and often attributed to the Red Queen but actually uttered by Humpty Dumpty:

> 'When I use a word', Humpty Dumpty said, in rather a scornful tone, 'it means just what I choose it to mean—neither more nor less'.
> 'The question is', said Alice, 'whether you can make words mean so many different things.'
> 'The question is', said Humpty Dumpty, 'which is to be master—that's all'.

But Mittermaier was not just playing with the notion of nominalism and was, in fact, developing a realist approach in which facts play a central role and he was not on the same wavelength as the 'dogmatic' economists who were content with abstractions rather than reality.

2. His approach to economics

But now I should turn to the central part of his contribution.

Much of his discussion turns around the possibility of an omniscient Laplacean view of the world in which there are, in reality, well-defined causal relations, and where prediction and explanation of past events are just mirrors of one another, there being no place for chance and coincidence. Economists' attempts to achieve this omniscience are well summed up by Sargent's remark about rational expectations: 'All agents inside the model, the econometrician, and God share the same model' (Sargent, 2005, p 3). To this could be added Mike Woodford's observation:

> It has been standard for at least the past three decades to use models in which not only does the model give a complete description

of a hypothetical world, and not only is this description one in which outcomes follow from rational behavior on the part of the decisionmakers in the model, but the decisionmakers in the model are assumed to understand the world in exactly the way it is represented in the model. (Woodford, 2012, p 2)

The modellers are thus attributing much greater cognitive power to those whom they model than to themselves. Woodford is clearly sceptical of these hypotheses and, like Mittermaier, does not see people that inhabit real economies as the omniscient individuals who are portrayed in standard macro-economic models.

3. Is rationality inherent, learned, or something that is not a characteristic of ordinary individuals?

The notion of rationality lies at the heart of much of Karl Mittermaier's thinking. He clearly thought that the basis for any economic theory should come from what he called *ex ante* facts rather than *ex post* facts. This distinction, which might puzzle some people, is, if I understand him correctly, based on the idea that there are some regularities in the process that governs the evolution of an economy but many of the data from the past are not of much use in helping us to understand the general structure of the economy and, even less, to be able to predict the future of that system. However, he shared Hayek's view that there are recognizable 'patterns' which recur, and which do help us to fathom the nature of economic activity. Yet he thought that the structure that economists have hung on the *ex post* data was not only unsatisfactory but probably wrong. To put things simply, suppose that we try to explain past developments in economic activity as being the result of individual optimization over the alternatives available to them. So, individual self-interested optimization would be the axiomatic basis for the analysis of economic outcomes. But why use the term 'axiomatic'. This comes from the desire of economists to place their discipline in a framework which could be treated mathematically, and within which one could 'prove' results.

The framework that emerged with the marginal revolution and honed to perfection by Arrow and Debreu (1954) was based on a series of assumptions that were certainly not drawn from reality and when examined closely, as Herb Simon was fond of pointing out, had little to do with how people actually behave. Mittermaier had little time for the assumptions normally made by economists about people's preferences. He had penetrating insights about the nature of those preferences and the space over which they were defined. His discussion of the consistency of preferences is particularly interesting.

4. *Ex post* and *ex ante* facts

As I have said, Mittermaier put great emphasis on 'facts', but the terms *ex ante* and *ex post* are likely to puzzle some. How can one see this distinction in economics? My view is that Mittermaier was expounding an idea that has been taking shape through a variety of channels but that, rather than modify some of the arguments that were out there, he constructed something which he found satisfactory, but which uses terms that can cause confusion. When he talks about the *ex post* facts that one uses to construct explanations or stories to explain the past, he clearly denies the idea that we can extrapolate from that set of facts to predict the future. Why is that? It is because what happened is not only what results from the deterministic effects from certain causes but is also conditioned by chance and coincidence. One cannot, he argued, extrapolate from these random occurrences. Hence the evolution of the economy is globally unpredictable. He asserts that, nevertheless, there are recurrent patterns and, in this, he joined Hayek who asserted that 'there are no laws in economics just patterns' and the job of the economist is to recognize those patterns. Mittermaier leaves us in a halfway house. The world, he says, is not deterministic and does not approach the physical sciences in that respect. However, the world is not totally without structure, and we should pragmatically seek to understand that structure. Econometrics seems to have pursued this route by postulating deterministic models which are subject to the influence of 'noise' where that noise is responsible for the 'deviations' from the deterministic path that the system would have followed. As soon as we accept this then we are on the road to adopting the structure that I have mentioned which can be thought of as a deterministic model which is persistent and is just perturbed by the noise or exogenous shocks.

However, as a number of people have pointed out, the underlying assumption which justifies this is what is called the 'ergodic hypothesis'. This says that a process is ergodic if the probability distribution of the states through which it passes over time converges to the probability distribution over the states in which it might find itself at any point in time. If you ran the process many times and, each time, recorded its state at one particular point at time the distribution that you would get would be the same as if you had run it for an infinite time.

It has frequently been observed that 'the world is not ergodic', and it has been argued that without this assumption prediction is not possible. Samuelson said, for example, that without the ergodic hypothesis economics becomes 'just history'. The argument about the importance of ergodicity in economics became quite fierce with interventions from O'Donnell (2014) and Davidson (2015) for example. The argument turned to a considerable extent as to what Keynes's opinion or position was on the nature of uncertainty. I have the impression that Mittermaier was walking gingerly around this idea but without directly referring to the notion of ergodicity.

5. Realism

When discussing the apparent failure of economists to take a realistic attitude to the economy that they are analysing Mittermaier quotes Eddington who chastises physicists for having been ensnared in the technicalities of their discipline as opposed to interesting themselves in the actual phenomena with which it should be concerned.

> This vagueness and inconsistency of the attitude of most physicists is largely due to a tendency to treat the mathematical development of a theory as the only part which deserves serious attention. But in physics everything depends on the insight with which the ideas are handled before they reach the mathematical stage. (Eddington, 1939, p 55)

Here one can think of the echoes from Marshall,[1] far from the approach that modern economic theory has taken. The use of mathematics should not, according to many mathematicians, be confined to looking around for some existing results which might be convenient to solve the problem at hand but should play a more creative role. To cite a very well-known mathematician, Sarah Hart,[2] the Gresham professor of mathematics:

> Real mathematics involves not knowing what is going on, not having any idea what to do, and then playing around and hopefully finding your way through. Finding the way often involves imposing structures and constraints on a problem. The tension is between wanting the most general result possible and actually being able to prove something. You could prove hundreds of rubbish theorems about your very precise special case, but nobody would care because it has no wider implications or applications. You want just enough structure to hang your ideas on, but not so much that you are boxed in. (Hart, 2021)

What she suggests is that, rather than imagining some abstract framework, what one should do is to look for some framework that has some basis in reality and to hang one's hat on that. I think that this would be much more in keeping with Mittermaier's vision than what he described as the 'dogmatic' approach. Indeed, I feel that he was looking for ways of making economic theory 'better' and that this could not be done by making small modifications to the model that we inherited from our enlightenment ancestors. Intellectually, I do not see him as a 'classical liberal' though, for all that I know, that expression might catch some of his general philosophy.

6. Preference fields

Among the building blocks of modern economic theory preferences play a central role and Mittermaier discusses their nature at length. Anyone who has looked at modern micro-economics knows that we impose extraordinary conditions on them, such as they are defined over a space of goods or services, for the most part taken to be Euclidean, and have the properties of reflexivity, transitivity, continuity, monotonicity, convexity and so forth. None of these assumptions would seem plausible to an uninformed bystander but one can look at it from the other side and observe that with all these assumptions one can derive a certain number of results. But they are, of course, at best some sort of abstract approximation which allows us to formalize the idea that individuals when they choose do maximize something.

I suspect that transitivity was about as far as Mittermaier wanted to go and that he regarded the rest as providing the structure for the maximization exercise that individuals are presumed to do. But, even if we did accept that people have preferences over the alternatives available to them, then we are not really requiring that they have them over all the possible alternatives that might be available to them. In fact, to do so makes the problem much more complicated.

It would entail dealing with an infinite number of goods unless we were prepared to endow people with a finite life. But to do the latter would also mean stipulating who is alive when and who are the others with whom exchanges and agreements can be made. As it is, as has often been said, the whole framework is set up at time zero and the world either unfolds in a purely deterministic way or the individuals make their decisions at time zero and the world evolves in a stochastic way, but a way which is mysteriously understood by the participants.

But if individuals are to make decisions, at the outset, they must know which alternatives are available to them. This means that there must be constraints. But where do those constraints come from? In the Walrasian model there is some actor who announces the rates at which units of commodities can be exchanged with each other, that is this actor announces prices which are observed by everybody. This actor is commonly referred to as the 'auctioneer' even though, as Walker (1996) has clearly shown, Walras never used the term. Once prices are announced, in this admittedly over-simplified model, people can then know the value of their own goods and this defines their budget or wealth, or income. The two being equivalent in this case. The individuals can then choose their actions and acquire and divest themselves of goods so as to obtain the most satisfying bundle to which they have access. When the prices are such that what is demanded of every good is just equal to what the participants in this economy are willing to supply, there is equilibrium. What I have just said involves major restrictions on the behaviour of the agents and on the institutional structure of the economy.

7. The problems

First of all, consider the fact that in any system in which agents interact, that interaction will influence people's choices and even a detailed study of an individual in isolation will tell us little about the aggregate outcomes. We are dealing with a complex system and, if there is one thing that has become clear, it is that one cannot simply derive aggregate behaviour from our knowledge of the individual components of the system.

Nobody believes that the framework that I have just outlined corresponds to any market that exists or has existed. The only reason that we have reached this point is our desire to construct a global model which can be solved, and which has to be a drastically simplified one to be solvable. But the easiest way to do this was to make the step from the individual to the aggregate directly whereas there were clear warnings that this was erroneous.

To cite Robert Laughlin, a Nobel laureate in physics:

> I am increasingly persuaded that all physical law that we know has collective origins, not just some of it. In other words, the distinction between fundamental laws and the laws descending from them is a myth. ... Physical law cannot generally be anticipated by pure thought, but must be discovered experimentally, because control of nature is achieved only when nature allows this through a principle of organisation. ... What physical science has to tell us is that the whole being more than the sum of its parts is not merely a concept but a physical phenomenon. Nature is regulated not only by a microscopic rule base, but by powerful and general principles of organisation. (Laughlin, 2005, Preface)

In a similar vein Phil Anderson, another Nobel laureate in physics, said:

> The ability to reduce everything to simple fundamental laws does not imply the ability to start from those laws and reconstruct the universe. In fact the more the elementary particle physicists tell us about the nature of the fundamental laws, the less relevance they seem to have to the very real problems of the rest of science much less to those of society. ... Instead, at each level of complexity entirely new properties appear and the understanding of the new behaviours requires research which I think is as fundamental in its nature as any other. (Anderson, 1972, p 393)

Where have we got to in our search for the overall model? Can we continue to satisfy ourselves with a framework that ignores the problem of aggregation. If, as these quotes suggest, aggregate outcomes are somehow intrinsically different from individual ones then much of standard economic theory built on 'sound micro-foundations' will not be satisfactory. This seems to be to be one of the major flaws of modern theory.

8. Coalitions and their role: an insight from game theory

The basic idea of the 'Invisible Hand', which is the focus of Mittermaier's other book, is that there is some mechanism which yields 'satisfactory' allocations of the resources with which the economy is endowed, and this mechanism is referred to, in a distortion of what Adam Smith actually said was 'The Invisible Hand'. The underlying reasoning is that a group of individuals acting in their own interests will self-organize into a state which has, from a global point of view, 'good properties'. As Mittermaier points out, what the set of possible outcomes is and the criteria for deciding which of them are desirable depends very much on the structure that one imposes on the economy. The easiest framework is that which has traditionally been used in the modern neo-classical picture of a market or economy and is what is referred to as an 'exchange economy' and Mittermaier uses it as a starting point for his discussion of what constitute satisfactory outcomes. In an exchange economy what is involved is a world in which there is a fixed number of goods and a fixed amount of each of them, which can be allocated among a fixed number of individuals each of whom has given preferences over bundles of these goods. The goods are initially held by the individuals and are referred to as their 'endowments'. Think of a marketplace in which producers of different goods bring these to the market to sell and wish also to buy other goods to satisfy their wants or needs. Some assumptions are made about the preferences of the participants in the market. For example, it is assumed that people prefer more to less of any good, that if they prefer bundle x to bundle y and bundle y to bundle z then they will prefer bundle x to bundle z, (transitivity) these assumptions are held to define what is 'rational'. The question, then, is which allocations in such a framework are satisfactory?

Within the simple example of an exchange economy there are two obvious candidates for the set of aggregate outcomes. One is the familiar idea of a Pareto optimum, an allocation of all the goods to the individuals where no reallocation of those goods could make somebody better off without making someone else worse off. There are many such outcomes, indeed, from several points of view, too many. Another is the so-called 'competitive equilibrium', which introduces the notion of price. Suppose that in the marketplace each good has a price and that everyone in the market knows those prices. Then, each participant can calculate the value of the goods they bring to the market, and this constitutes their wealth, often referred to as their 'income'. Then the individual knows that they can purchase any bundle of goods which has a value less than that of their income which they will spend in return. If prices are such that the total quantities of the goods demanded (aggregate demand) is just equal to the quantities of the goods brought to the market (aggregate supply), then the result is a competitive equilibrium. We can easily show that a competitive equilibrium is a Pareto optimal allocation, and it is, of course, one that makes some sense from an economic point of view.

However, finding a restrictive set of allocations which had satisfactory properties was not useful for answering the basic problem that Mittermaier posed. Even if the allocations mentioned might be satisfactory, what is important is to know how they might be attained. Mittermaier wanted to find an approach which would provide an answer to that question. He was convinced that arriving at reasonable allocations of goods involved groups with particular affinities. In other words, it is not just a mechanical allocation depending on some prices determined by some anonymous actor but rather an agreement among like-minded individuals which sustains the arrangement. It is intriguing that Mittermaier, in this search for ways of defining satisfactory outcomes for an economy, should have hit upon the notion of the core. This is a 'solution concept' which gave rise to a large but very abstract literature. Given his background and his philosophical bent it might seem odd that a concept from game theory would come to his mind. Perhaps even more so because one should always keep in mind that when people are talking about this type of allocation they are typically talking about a purely static concept. Yet, Mittermaier was also looking for some sort of procedure which would lead to the acceptable allocations. It is here, in my view, that Mittermaier showed that he had not only found allocations which were satisfactory from the global point of view but ones which could be thought of as emerging from collective behaviour.

To explain, let me first describe what is meant by a core allocation. Think of any arbitrary allocation of goods to the participants in the simple exchange economy I have already described. Now this will specify which 'bundles' of goods will be received by which of the participants. This is, as if all the endowments of the community in question were put together in one place and then someone would suggest a reallocation of those goods among all the participants. Then the question is asked, 'Does any individual or group have an objection to this allocation?'. By this is meant, is there any individual or group who could withdraw their initial endowments from the collective heap and then redistribute them either to him or herself or to the members of the group so that they are made better off than in the original allocation proposed? If this is the case, then the individual or group in question can make an objection to the allocation proposed. Any allocation to which no individual or group has an objection is in the core. In the literature that developed around this notion it was originally assumed that any coalition could object to (or 'block') a proposed allocation. This meant that any allocation in the core is Pareto optimal since if this were not the case the coalition of the whole community could block it and improve everyone's welfare. Furthermore, it is easy to show that any competitive equilibrium allocation is also in the core (see Hildenbrand and Kirman, 1988). This is interesting because no mention is made of prices in defining the core. A standard objection to the usual Walrasian equilibrium is that there must be someone who calls out the prices which are then taken as given by all the market participants, and the typical explanation is that there is a 'commissaire priseur' or auctioneer who does this. So, this is far from the usual characterization of a perfectly competitive Walrasian market as one governed by a decentralized procedure.

What stimulated so much interest in the core was that, although it involved no mention of prices, it was possible to show that in large economies, in terms of the number of agents the only allocations in the core were the competitive equilibrium allocations. So, an increase in the number of economic agents made the core 'shrink' to the set of competitive equilibrium allocations without prices ever entering the scene.[3]

But – and here I see evidence of a remarkable insight of Mittermaier – he contemplated how the core allocations might actually be generated by some collective arrangement. He thought that it was unreasonable to think in terms of arbitrary coalitions as removing unsatisfactory allocations but rather thought that one should only consider coalitions which might have a realistic possibility of forming. Several of us played with the idea of putting restrictions on the coalitions that could appear, but as in Kirman et al (1986), we thought of individuals as being in a network and only allowed them to form coalitions if they were close to each other. For example, one could limit the coalitions to groups of individuals all of whom were directly linked to one another. Or could require that for any two people to be in the coalition, they must both be linked not necessarily to each other but to someone else in the coalition, that is they should have a 'common friend' in the coalition. But this is still very abstract, for what is not described is the nature of the links and what is necessary, as Mittermaier realized, is that the underlying network should have some sociological or economic reasoning for existing. For this we can exploit our knowledge of people's family, religious or other ties. It is, of course, much more plausible to approach the problem like this rather than to assume an arbitrary network. Such an idea would be consistent with some of the Austrian writers in whom Mittermaier showed such interest. But one could also trace such an idea to Pareto with his theory of the 'Elites' who, when they were in power, would eliminate allocations of resources which did not favour themselves.

To take Mittermaier's idea seriously we should specify what is meant by closeness, that is, we should define a metric over the set of all the participants in the economy or market we are discussing. The usual metric in a graph is the number of links between two participants; the measure of how well connected the whole graph is, is that one takes the number of links between the two participants that are the least well connected. This is the so-called diameter of the graph. Now suppose that we do not know the structure of the whole graph, but do know the probability that any two participants will be linked to each other. So, the relations between individuals are stochastic. This is much more plausible than the idea that we know the details of all the links. As we look at larger and larger graphs the probability that any two individuals will be linked should decrease. A remarkable result of Bollobás (1985) shows that if this probability does not go to zero too fast as n the number of participants in the group increases then finally the diameter of the group will be 2 with probability 1. In other words, every pair of individuals in that group will have a common friend who is in the group. Thus, a very large coalition will be very tightly linked. This is counter-intuitive since when two people meet who do not know each other but find out that they have a common friend, their usual reaction is to say, "Oh, it's a small world", whereas they should conclude the opposite, that the world is large. Now how does all of this link with Mittermaier's ideas about coalitions?

Well, what is being discussed is what defines or characterizes a coalition and the idea is that its members have something in common which links them together but using the word 'coalition' does not avoid the problem, for in the way Mittermaier was thinking of a coalition, it was less some common characteristic but rather some common purpose that brings the members together. They get together because it is in their common interest to do so.

But here we are back to the nominalist debate. Calling them a coalition rather than a group or family does not change that.[4] As Stettler (2019) in his analysis of Mittermaier's philosophy points out, Darwin took this 'nominalist' position when he talked about species, arguing that the actual term had no significance without reference to some physical property. Pursuing that line of thought one can interpret the expression 'it does not matter' in a similar way. Or to go further as Oscar Wilde said, 'Nothing matters very much and most things don't matter at all.'

But why was this digression about the core and its properties justified. For me, this is a perfect example of Mittermaier's approach. Although he was clearly influenced by those whom he met or read he was an essentially independent thinker and one who followed a trail where it led. So, in arguing that the use of a game theoretic notion could help to fill one of the most important gaps in economic theory, he was 'thinking out of the box'.

He seems to have had in mind that the common interests of people in a coalition, or whatever one would like to call it, would lead them to reject material arrangements which could be changed to improve their collective welfare. How this would happen and what institutional arrangement would facilitate this is not stated. However, Mittermaier was evidently trying to find a way in which the satisfactory outcomes that the Invisible Hand was supposed to produce could actually be generated. The appeal to the idea that individuals with common characteristics or goals might act together to obtain what they want or, at least, to prevent things that they did not want does not seem unreasonable. It remains to be specified through what social process this would happen. Here Mittermaier was pursuing one of his important themes, the notion that people's preferences are strongly conditioned by the institutions of the society in which they exist. Again, as he points out, the term institution can have many meanings.

In one sense the idea of using the core or some other solution from cooperative game theory is a major step forward, from some imaginary mechanism in which people somehow send signals and receive messages and these are adjusted by some central mechanism until equilibrium is reached. The problem with the mechanism design approach is that the mechanisms do not emerge naturally from the interaction of agents but provide a carefully constructed framework within which the participants have to work. Whereas it seems to me that Mittermaier seems to suggest a process rather like the Iowa caucuses. But once again the rules have to be specified at some point though one could think of an iterative process of negotiation. But, to come back to the basic point, I see Mittermaier's approach as a significant step in incorporating the links that exist between individuals into the process of collective decision-making.

9. An alternative route

Having taken a long detour down one of the interesting paths that Mittermaier hit upon, I would like to make a quick reference to an idea that, I think, would have pleased him. When Werner Hildenbrand became disillusioned with the road that General Equilibrium had taken he basically abandoned the idea of using individual optimization as a basis for economic theory. He argued in his book *Market Demand* (Hildenbrand, 1994) that one could show that well-known relations in economics, often referred to as 'laws', could be derived from the choices that people actually made without specifying how or why they had made those choices. He did this for the 'Law of Demand' to which Mittermaier often referred. The idea is simple. In the simplest case, that of a single commodity, it simply says that when the price of a good decreases more of that good is purchased. This is easily generalized to the case of l goods and in that case says that the price vector moves in the opposite direction to the goods vector. In the single good case we say that the demand curve is downward sloping. But what is chosen in aggregate need not be demand derived from individuals' preferences. We just need to know the quantities of goods that were purchased at different prices. However, to derive the aggregate law of demand we have to make some assumption about the nature of the choices. What Hildenbrand did was to assume that as prices rose the distribution of the amounts chosen by individuals became more spread out. Obviously this has to be correctly specified and tested empirically but from this assumption about the choices of all individuals one can prove the 'Aggregate Law of Demand' without having to reflect on the motives that people had for making those choices. It avoids so many of the restrictive assumptions we have made about people's preferences and is completely rigorous. Yet it gained little traction in the economics profession. I suspect that Mittermaier would have enjoyed thinking about it.

10. Conclusion

This book reveals a fascinating personality, widely read, influenced along his path by a variety of different people, various representatives of the Austrian school, but also taking a careful look at the formalists, many of whom are well described as 'dogmatic'. His strength was more logical and philosophical than mathematical, but this did not prevent him from pointing out some of the weaknesses in modern theory, which we too often skate over as quickly as possible. We have good reason to do so because otherwise we are likely to fall into rather cold water. I can only repeat that I regret not having had the opportunity to meet with Mittermaier and was struck by the many points that he focused on and which have given me pause for thought over time. I have probably spent too much time on the subjects that he dealt with which I am familiar with and may not have done him justice for the rest but even within that narrower field I learned a considerable amount from him.

Centre d'analyse et de mathématique sociales (CAMS)
at École des Hautes Études en Sciences Sociales (EHESS), Paris

Notes

[1] Mittermaier discusses in this book the stand that Marshall took and Friedman's (over-) simplification of it. Marshall at one point argues that one may use mathematics to clarify one's statements but once one has reached a coherent conclusion one should throw the mathematics away. Not an attitude that is adopted by most authors in the leading economic journals.

[2] Sarah Hart is professor of geometry at Gresham College in London where she holds the oldest mathematical chair in England, established in 1597, and is the first woman to do so.

[3] The first theorem in this direction was proved by Debreu and Scarf (1963), although I have seen nothing in Mittermaier's work to suggest that he was aware of it and indeed his interest lay elsewhere.

[4] Rob Axtell and Doyne Farmer (2023) refer to the literature on coalition formation, and say that such models as have appeared in economics have been based on the idea that the members of a coalition have some characteristic in common, but do not go on to argue that such coalitions play an important part in macro-economics.

References

Anderson, P.W. (1972) 'More is Different', *Science*, 177(4047): 393–6.

Arrow, K.J. and Debreu, G. (1954) 'Existence of an Equilibrium for a Competitive Economy', *Econometrica*, 22(3): 265–90.

Axtell, R.L. and Farmer, J.D. (2023) 'Agent-based Modeling in Economics and Finance: Past, Present, and Future', *Journal of Economic Literature* (forthcoming).

Bollobás, B. (1985) *Random Graphs*. London: Academic Press.

Davidson, P. (2015) 'A Rejoinder to O'Donnell's Critique of the Ergodic/Nonergodic Explanation of Keynes's Concept of Uncertainty', *Journal of Post Keynesian Economics*, 38(1): 1–18.

Debreu, G. and Scarf, H. (1963) 'A Limit Theorem on the Core of an Economy', *International Economic Review*, 4(3): 235–46.

Eddington, A. (1939) *The Philosophy of Physical Science*. Cambridge: Cambridge University Press.

Hart, S. (2021) 'Triangulating Math, Mozart and "Moby-Dick"', interviewed by Siobhan Roberts, published 6 March 2021, updated 9 March 2021, *New York Times*, https://www.nytimes.com/2021/03/06/science/math-gresham-sarah-hart.html.

Hildenbrand, W. (1994) *Market Demand: Theory and Empirical Evidence*. Princeton: Princeton University Press.

Hildenbrand, W. and Kirman, A.P. (1988) *Equilibrium Analysis: Variations on Themes by Edgeworth and Walras*. Amsterdam: North-Holland.

Kirman, A., Oddou, C. and Weber, S. (1986) 'Stochastic Communication and Coalition Formation', *Econometrica*, 54: 129–38.

Laughlin, R. (2005) *A Different Universe*. New York: Basic Books.

O'Donnell, R. (2014) 'A Critique of the Ergodic/Nonergodic Approach to Uncertainty', *Journal of Post Keynesian Economics*, 37(2): 187–209.

O'Donnell, R. (2020) 'Karl Mittermaier, a Philosopher-Economist with a Penetrating Intellect and Twinkling Eye' in K. Mittermaier (2020) *The Hand behind Invisible Hand: Dogmatic and Pragmatic Views on Free Markets and the State of Economic Theory*. Bristol: Bristol University Press, pp 235–55.

Pareto, V. (1935) *The Mind and Society: Vol 1*, translated from the Italian *Trattato di Sociologia Generale* by A. Bongiorno and A. Livingston. New York: Harcourt Brace.

Sargent, T. (2005) 'An Interview with Thomas J. Sargent', interviewed by G.W. Evans and S. Honkapohja, CESifo Working Paper No. 1434.

Stettler, M. (2019) 'An Introduction to Karl Mittermaier and His Philosophy of Economics', *Journal of Contextual Economics*, 139: 123–34.

Walker, D.A. (1996) *Walras's Market Models*. Cambridge: Cambridge University Press.

Woodford, M. (2012) 'What's Wrong with Economic Models?', https://www.ineteconomics.org/research/research-papers/whats-wrong-with-economic-models.

Prologue: Mittermaier's Conceptual Framework

Rod O'Donnell

In 1976, nearly 50 years ago, Karl Mittermaier submitted an unsupervised dissertation to an economics professor. It received a poor reception, not because it lacked merit but because it was not understood. In the absence of feedback or suggestions, it was put aside.[1] That it was not understood is unsurprising for it covers a wide terrain, approaches economics from a philosophical angle, introduces novel concepts, presents a deep and reflective work, and requires time for the arguments and their interconnections to be appreciated. Thanks to the tireless efforts of his widow, Isabella, recommendations from readers, and Bristol University Press, it is now publicly available.

His setback was both our gain and our loss. Our benefit was having two major works from his pen, one pursuing discipline-wide foundational matters, and the other analysing foundational matters in Smith and Smithian exegesis. Our loss was that the first was stillborn, for had a degree been awarded at that time we would doubtless have had more contributions from his pen on a most important subject.[2]

Unlike modern dissertations, however, it does not begin with introductory remarks outlining its aims, main ideas, methodology or conclusions. It dives straight into its subject matter, and only partway through the first chapter do we learn of its grand objective. Similarly, the final chapter provides neither synthesis nor summary, but focuses entirely on the last element in his framework.

It was thus suggested that a prologue providing a brief overview might assist initial understandings, encourage deeper understandings and promote further discussion of this complex, subtle and carefully worded work. Because brevity can interfere with accuracy, what follows is a sketch-map, not a detailed representation of the territory traversed, a partial resumé of certain key issues rather than a complete account. The economic schools discussed largely concern neo-classical and Austrian economics which, for conciseness, are sometimes here termed orthodox economics due to their similarities rather than their differences.[3]

1. A revolutionary goal, modestly presented

Mittermaier's reading in economics, philosophy and history left him highly dissatisfied with the current state of economics and motivated his efforts to improve it as a scientific discipline. Of the two options available – retain and revise, or reject and replace – he chose the more radical course. And, although not underlined at the start, the construction of a new conceptual framework for economic analysis to replace dominant orthodox frameworks is its grand objective. The 'conventional' frameworks need to be abandoned, and a better alternative devised.[4]

2. Primary components

His new framework is founded on a realist philosophy valorizing empirical and historical observation, not an idealist philosophy valorizing non-empirical, timeless or empty abstractions. All its elements are interrelated and realism-based, with the following playing central roles.[5]

1. Facts, and their different kinds.
2. Induction as a means of establishing facts and axioms for deductive theorizing.
3. Institutions.
4. Historical time.
5. Determinacy and indeterminacy in outcomes.
6. Propositional logic, not purely mathematical logic.

He described his new framework as 'obvious', but with this obviousness obscured by the nature of orthodox economics.

3. Facts

Facts are vital to his analytical enterprise, with two distinctions drawn.

3.1 Ex ante *and* ex post *facts*

At first sight, this appears odd. Surely no fact could be *ex ante*, for facts refer to events that have happened or are happening, as in 'it rained' or 'it is raining'. His meaning is different and novel, with subtle relations to temporality.

Ex post facts refer to events that have occurred in the past prior to *some* moment in time. The subtlety is that this moment can either be now (in which case the facts have happened and are knowable), or in the future (in which case the facts are similar to predictions with their truth known only after the future moment has passed). In both cases *ex post* facts refer to what happened once we reach a given moment, and hence to what then lies behind us at that moment.

Ex ante facts refer to facts that exist before *ex post* facts occur. The most important kind are *the structural features of the economy that generate the ex post facts as outcomes*. In this sense, *ex ante* facts are causes, and *ex post* facts are effects. An economy is viewed as having an empirical, factual structure described by such matters as its institutions and their interrelationships, the agents performing various roles within the structure (either individually or in combination), and the objectives and capacities of the agents in pursuing their goals. In combination, these structural features (or *ex ante* facts) cause certain outcomes to occur. Hence they play central roles in scientific theorizing that seeks to understand the economy, and explain why it behaves as it does, as indicated by outcomes in the past (or outcomes yet to be revealed in the future). These outcomes are the *ex post* facts caused either fully by the *ex ante* facts or largely by them (see section 4).

Two related distinctions are involved. One concerns answers to questions. *Ex post* facts answer backward-looking questions such as 'what happened?', and make 'That X' statements, as in 'It is the case that X happened'. By contrast, *ex ante* facts provide answers to 'why did that happen?', and so deliver explanatory responses to 'Why X?' statements, as in 'This is why X occurred'. The second distinction is between explanation and prediction. Here *ex ante* facts not only play major roles in explaining the *ex post* facts of the past, but also in predicting the *ex post* facts possibly occurring in the future.

Obviously, Mittermaier's distinction does not refer to the *same* specific fact at different times, say today's prediction of tomorrow's weather compared to what tomorrow actually delivers. It is about 'two entirely different orders of fact'. The *ex ante* facts refer collectively to the structural features of the economy that generate the specific *ex post* facts that will occur if these causal structures have been correctly identified, remain constant, and nothing else intervenes. The temporal or before-and-after aspect of the distinction is associated with theorization in two senses: causality, because causes precede effects; and theoretical reasoning, because (a) premises precede conclusions, and (b) conceptual frameworks precede the specific theories developed therein.

3.2 Causal and casual facts

Despite similarity in spelling, these must not be confused. Causal facts are the *ex ante* or structural facts present in the causal chains determining the *ex post* facts, *ceteris paribus*. Casual facts are time- or place-specific facts arising 'accidentally', not structurally. In general, observed *ex post* outcomes are generated either by causal facts alone or by combinations of causal and casual facts.

Physical mechanisms provide examples. In an automated production line, if all machines are well-made and maintained, and no external interferences occur, the output is generated only by the causal facts and will satisfy prescribed tolerances. But if such provisos are not met (say due to poor machine manufacture, excessively worn parts, or the presence of dirt), the output will be created by a combination of causal and casual facts, and may be unsatisfactory.[6] Most of Mittermaier's discussion focuses on causal facts.[7]

3.3 Confusing ex ante *and* ex post *facts*

This major criticism is levelled at orthodox theorizing. Whatever the observed *ex post* outcomes might be, they are seen as identical to the outcomes delivered by the *ex ante* 'facts' of its theory. If theory concludes that well-organized free market systems generate universal optimization, then that outcome is what will happen in reality (with or without short adjustment times).[8] The outcomes deduced from the theory's *ex ante* facts are seen as replicated by the *ex post* facts of reality.

Two implications follow. First, explanation and prediction become identical. The same theory is deployed but in reverse directions: explanation is backward-looking and prediction forward-looking just as in 'classical mechanics'. Here the argument is that good economic theorizing separates the two activities. Good explanations of the past are always available when theories have realistic axioms, but fully reliable predictions of the future are rarely, if ever, available from any economic theory. The second implication concerns determinism and indeterminism. Orthodoxy delivers deterministic equilibrium outcomes, not indeterministic ones. The former arise from idealized axioms (say perfect knowledge and abilities in the neo-classical case) chosen to deliver the best possible outcomes (universal optimization). By contrast, non-orthodox frameworks are capable of delivering non-optimal equilibrium outcomes.

4. Induction

How do we obtain knowledge of the *ex ante* facts or structural features? By observing reality and drawing inferences. Induction then becomes central to the new framework because it is the key means of gaining an understanding of the elements playing causal roles in determining outcomes – institutions and their properties, uncertainty in various forms, the scope and effectiveness of our capabilities in decision-making, and the mechanisms generating outcomes and any variabilities in these outcomes. The common or repeated features of the *ex post* facts thus lead to the *ex ante* facts or structural features generating the observed outcomes.[9] Logical reasoning based on these facts then delivers conceptual frameworks and hence the theories inhabiting these frameworks.

Induction offers no support whatsoever to the perfections informing the axioms of neo-classicism – agent omniscience (including preference orderings over vast numbers of possibilities, and uncertainty-absence), perfect calculating abilities, and perfectly competitive markets generating universal optimization. Observation of real humans and markets will never inductively generate propositions about the imaginary perfect beings, capacities and outcomes of idealism-based theorizing. The facts entering genuine inductions are the facts presented by reality, not imaginary facts or fictional reconstructions of actual facts. The inputs are based on observed past facts and the outputs are observable future facts. If future facts differ from past facts, revised inductive conclusions will emerge (given the absence of casual facts).

Induction has been criticized, from at least Hume onwards, as deductively unjustifiable. But from the 1950s at least, philosophers have advanced theories providing non-deductive justifications of induction. These deliver philosophical underpinnings for both rational beliefs based on existing data, *and* rational changes in rational belief when justified by additional data. Deduction using axioms assumed to be universally true eliminates doubt in conclusions, whereas deductions based on inductively derived axioms always involve the possibility of doubt for induction never eliminates doubt entirely.

5. Institutions

The dissertation begins and ends with this core element of all real economies. Chapter 1 opens by claiming that his approach differs from that of early institutionalism by seeking a primarily analytical, rather than a primarily historical, approach. Whether one agrees or disagrees with his alternative approach and its accompanying definition of institutions, the key points are twofold: institutions play central roles in the new framework (as against orthodoxy which omits or sidelines them due to its foundations in individualism); and institutions are understood as *ex ante* facts playing analytical roles in economic theory. The final chapter returns to institutions as components of historical evolution.

6. Genetic understanding

This idea (also called narrative understanding) draws on the concept of 'genetic explanation' to provide accounts of how a 'system has developed into its current form from some earlier stage'.[10] Mittermaier's reframing as 'genetic understanding' emphasizes its links to '*Verstehen*', or intuitive understandings of how events involving humans can occur as they do, for we, as current humans, can understand the situations facing past humans and the decisions they made. The idea explicitly involves history, causality and change whether relevant to large systems (economies), medium systems (institutions), and very small systems (inter–individual negotiations).[11]

In Mittermaier's framework, temporality is explicitly present from the start, being inherent in the distinctions between *ex ante* and *ex post*, cause and effect, explanation and prediction, and possible revisions of *ex ante* facts. The economy, its institutions and its agents all move through time from past to present to future, with significant openness present at the *beginning* of analysis so that we understand how the present developed from the past and can influence the future. That must not be ignored by using abstract universal constructs eliminating or trivializing temporal change.

In economics the difference is between empirical constructs drawn from reality and non-empirical constructs drawn from pre-conceived abstractions in idealized realities. In the former, reality figures in the construction of its *explanans*, while in the latter reality only enters via particular interpretations of the *explanans* and *explananda*. Where the former is open and emerges from the evidence, the latter is closed and imposes itself on the evidence to ensure consistency with pre-given conclusions. Otherwise put, the difference is that between allowing the system *itself* to determine the possible outcomes of its multiple interdependent elements, versus insisting in advance that the system always reaches the same destination of universal agent optimization.

This difference is closely related to other key elements in Mittermaier's framework. One is the *ex ante/ex post* fact distinction. Genetic understanding provides explanations that keep these separate; *ex ante* facts exist before, and help explain, the *ex post* facts that occur in some future, *ceteris paribus*. *A priori* deductive explanations, however, equate them. The *ex ante* facts presumed to exist (in keeping with its axioms) become identical to the *ex post* facts that occur in reality. A second concerns induction. The *ex post* facts of the past contributed to the *ex ante* facts of the present, but the *ex post* facts arriving in future may differ from those of the past, so leading to revised inductive conclusions and hence revised *ex ante* facts. Genetic understanding allows difference between the two kinds of facts, whereas explanations based on universal propositions do not. Finally, there is the important issue of whether a theory embraces determinism (in the sense of always reaching the same unique outcome from the universal axiomatic constructs in an idealized world), or allows for non-determinism (as in beginning with empirically based constructs drawn from the real world that allow for variability in outcomes across time and space).[12]

7. Logically valid conceptual argument

Once *ex ante* facts have been inductively determined, Mittermaier deploys conceptual deductive logic in his theorizing, not mathematical logic. The following are emphasized:

1. The understanding of meanings, as in 'the prediction of a chance event' is a 'contradiction in terms'.
2. The importance of drawing conceptual distinctions in economics and philosophy, and avoiding conflations. This constant theme in his thought is illustrated by his observation that 'logical consistency' differs from 'consistency over time', for the former has a separate meaning in axiomatic constructs due to 'the temporal order of experience not entering the picture'.
3. Obedience to the law of non-contradiction, and hence the avoidance of 'irreconcilable basic premises', at all times.

Overall, his conclusion is that two things are required in constructing the new conceptual framework: 'a logical reconstruction of the way knowledge is built upon knowledge', and 'a criterion for what sort of knowledge we can begin with'. *Inter alia*, the first involves certain key differences between propositional and mathematical reasoning, while the second involves differences between realism and idealism.

7.1 A methodology for scientific reasoning

A basic question is where to start in economics when developing frameworks and theories. Two main alternatives exist.

1. Begin with ideas drawn from reality, form axioms consistent with reality and develop theories capable of explaining reality. The concrete *explanandum* and the abstract *explanans* concern the *same reality.*
2. Begin with ideas drawn from some imaginary reality (say involving perfections and universality), form axioms based on that reality, develop theories consistent with this construct, and only later deal with any problems that arise. The concrete *explanandum* and the abstract *explanans* now concern very *different* realities. The former is real and imperfect while the latter is imaginary and perfect, so that bridging the gap then poses significant theoretical problems.

Mittermaier chose the former. The *ex ante* facts on which scientific conceptual frameworks must be grounded are the facts generated by the observed reality, not by sets of imaginary perfections thought to inform some meta-reality inside or beyond the world we actually experience, and which we would observe if only we could strip away the (largely human) imperfections delivering the empirical reality we do experience.

Consider, for example, his discussion of consumption. Realism-based analysts talk about 'things that everyone knows something about and can therefore criticize'. But orthodox preference field analysts only talk about 'a catch-all which no-one has experienced', and it is in 'the nature of a catch-all that it can explain everything and nothing'. The former supplies a realism-based explanation that may be acceptable or unacceptable on given criteria. The latter supplies *neither* a realism-based explanation, *nor* a logically acceptable explanation, for it is consistent with both everything and nothing. One might add that since this property is what contradictions alone can do, the catch-all will contain at least one contradiction. Mittermaier's softly worded criticism effectively says that neo-classical preference fields are inherently self-contradictory.

Another illustration concerns the current Walrasian conceptions of *all* market economies and Mittermaier's conception of *current* market economies. In Walrasian economies, the structural features (axioms) posit that all agents are self-focused utility maximizers, possess perfect knowledge (of commodities, contingencies and personal preference orderings), have perfect calculating abilities, and then have all their decisions coordinated by a time-zero auction in which the entire future is embraced, no trading occurs until universal optimization has been achieved, and no further economic decision-making occurs after time-zero. Here the assumed *ex ante* features of the theory (universal optimization and hence no unemployed resources) are *guaranteed* to be the *ex post* features of reality (universal optimization and no unemployed resources).[13]

By contrast, the *ex ante* facts in Mittermaier's framework pertain to reality. They currently concern a capitalist economy populated by agents who perform different roles, have imperfect abilities and knowledge, operate with expected values in an uncertain world, undertake transactions through historical time, make mistakes, often fail to meet goals, often operate in non-clearing markets, and inhabit a world in which outcomes can fall short of universal optimization. Here the *ex ante* facts allow a wide range of *ex post* facts. Equilibrium outcomes are not unique states or magnitudes, but involve a spectrum of possible states or magnitudes. Similarly, explanation and prediction are not identical. Good explanations of the *ex post* facts of the past are always possible, but providing good (reliable) predictions of the facts that will occur in the future is impossible because the mechanisms creating outcomes can generate multiple possible outcomes, and hence indeterminacy as to which particular outcome will appear.

8. Extending the framework

Mittermaier's framework is open to further development in multiple directions of which only three are noted here. First, expansion via extension or addition. Uncertainty is an enduring human reality and hence a major *ex ante* fact that can take various forms (such as probabilistic and non-probabilistic, for example), and possesses links to induction, indeterminacy, history and genetic understanding. Adding macro-economics to the micro-economics that is Mittermaier's primary focus is also possible.

Second, syntheses with other non-orthodox frameworks. Realism gives his framework (with or without modification) strong interconnectivity with other heterodox schools of thought such as institutionalism, Keynes's mature economics, behavioural economics, ecological economics and feminist economics. Finally, post-1976 contributions in economics and philosophy may be deployed to update, revise and expand the approaches taken to the key components, a move consistent with the historical development of scientific work in research programmes.

9. Conclusion

This rich, thought-provoking work contains more of economic and philosophical interest than has been canvassed here. On any subject, Mittermaier made valuable and penetrating contributions, with few things more important than ways to improve the methodologies, theories and policies of economics. He opened by remarking that 'many voices of dissent have been raised against economics', and closed by contrasting the inadequate explanatory power of *imaginary ex ante* facts with the superior power of *realism-based ex ante* facts.

If, despite possible differences in our backgrounds, we all share the goal of significantly improving economics, we need openness to better conceptual theorizing of this world, not more purely mathematized analyses of imaginary worlds. We should be guided by the same peaks that guided Mittermaier, Mt Improvement, Mt Realism, Mt Logic and Mt Fearless. Despite inadequate institutional support early in his academic career, the legacy of this deeply thoughtful philosopher-economist may still promote the cause he pursued. If the social science of economics is to deliver relevant and valid *explanans* for its *explananda*, and avoid fictional accounts based on imaginary worlds, mathematics-devotion or ideology, major changes are necessary.

University of Sydney

Notes

[1] One might extend sympathy to the readers who did not understand it, but no further. Later, in 1987, he obtained a doctorate with a brilliant dissertation on the more specific topic of Adam Smith; see Mittermaier (2020).

[2] Both dissertations are seminal contributions seeking improved understandings of their subject matters. While the second is easier to read, the first has greater relevance to progress in economics.

[3] Walras, Pareto, Menger and Marshall are key figures, for example.

[4] He sometimes called the latter a 'sceptic' framework, but this is insufficiently descriptive. What is advanced is not merely an expression of doubt, but an alternative. Given his student status, and the likelihood of orthodox examiners, his terminology was probably the wiser course: legitimate well-argued scepticism had more chance of success than calls for major change even if well-conceived and argued.

[5] As no one at the university kept a copy, the title of the submitted 1976 dissertation remains unclear. The present title is representative of its overall nature.

[6] Nagel (1961: 560n8) uses a gun and bullets.

[7] The distinction can also be related to *ceteris paribus* clauses. Note also that while casual facts are *ex ante* in time, they are not *ex ante* facts for they are accidental, not structural.

[8] 'Well-organized' means no interference from forces defined as non-market forces.

[9] Careful analyses of any differences between past and present inductive conclusions also assist in deciding whether these differences might be due to casual facts.

[10] See Nagel (1961: 20, 25–6, 551–75).

[11] *Verstehen* also connects humans as participants in economies to humans as theorizers of economies, an important point in Mittermaier's framework.

[12] From a scientific, realist and causal perspective, the proper handling of historical events and time are vital issues in economics.

[13] Rational expectations versions of neo-classicism employ similar assumptions except for the treatment of coordination and time.

References

Mittermaier, K. (2020) *The Hand behind the Invisible Hand*. Bristol: Bristol University Press.

Nagel, E. (1961) *The Structure of Science*. London: Routledge and Kegan Paul.

Institutions and the Empirical Content of Economics

1.1 The question of institutions

Over the years many voices of dissent have been raised against economics and the earlier political economy for seemingly failing to pay due attention to the institutions of specific societies and hence to the actualities of economic life. When critics, from Richard Jones to J.K. Galbraith, have produced works of their own which in their estimation have been free of this defect, they have never made more than a very modest impression on the mainstream of economic theory. Sometimes, as in the case of the German historical economists and to a lesser extent in that of the American institutionalists, these critics gathered for a time a following of their own and, among other things, contributed to the use of economic statistics and advanced the cause of social security legislation. But a single discipline which does justice both to the logic of economic relations and to empirical accounts of the institutional framework remained, except to convinced Marxists, an elusive ideal. The whole issue was debated impassionedly in the lengthy '*Methodenstreit*' begun by Menger and Schmoller, but no consensus was reached. There has remained to the present day what Eucken called the great antinomy between the individual-historical and general-theoretical approaches.

It may not always be easy to see what is at issue. Theory, it may be said, is surely not an end in itself, but the means for gaining a better understanding of a country's economy and its problems and, where applicable, for formulating appropriate policies. The institutional peculiarities of a country do enter this analysis as factors that qualify the conclusion. Theory may even be used to study the evolution of an institution, so that theory is the tool and the understanding of an institution the end product. In either case there is no conflict between the use of theory and the recognition of institutions. This is, very broadly, the way Menger presented his case in the *Methodenstreit*. Yet it does not really get at the heart of the matter.

One could perhaps interpret the institutionalist position as follows. Economic theories in themselves presuppose certain institutions (though Menger possibly thought that the 'exact' discipline of economics did not). Ricardian political economy, for instance, required a certain system of land tenure and a society whose members fell naturally into three categories. Modern general equilibrium theories require markets, a law of contract, private property, and so on, and perpetrate, according to some views, a subtle sin of omission in failing to recognize that tastes or preferences arise out of the evolving institutional structure. Now, the institutionalist critique has pertained to these presuppositions or implicit institutions and has wanted to see them replaced by the explicitly recognized institutions – those that merely qualify the conclusions of analysis or are the end products of analysis. It is the integration of theory and institutional fact that has not been accomplished, though some would claim that it was precisely Marx's achievement to have done so. Most theorists, it seems, are not opposed in principle to such an integration. They simply do not know how to bring it about. Some attempts in that direction have of course been made. The theories of monopoly and monopolistic competition are examples, as are the many variants in which equilibrium theory may be had with or without forward markets, information costs, and so on. But this falls very far short of what institutionalists have considered necessary.

However, the interpretation of institutionalism given here is not really in the spirit in which most of the critics in question wrote. In fact, they showed so little uniformity that it is difficult to generalize. The only traceable movement of this critical thought started with the historical school in Germany and eventually spread rather weakly to Britain and much more strongly, in the form of institutionalism, to the United States. Common to all in this movement, as it is to modern institutionalists, is a predisposition to see economic questions, in the way of 19th-century thought, in terms of cultural evolution and development. But beyond this, and even in this, there were great variations. Roscher and Hildebrand played with the idea of laws or stages of economic development, but Knies played down the expectation of finding laws in social evolution. Under Schmoller's guidance the movement in Germany turned towards a purer economic history, though the idea of finding common threads was not given up. It also influenced Max Weber's sophisticated historical analysis of institutions, but Weber was also part of a broader movement which included, among others, Dilthey and Croce. In England (more strictly, Ireland), Leslie made the emphasis on history and evolution into the injunction: 'Back to Adam Smith'. He and Ingram also brought the ideas of Comte into the subject. In the United States, the movement took the quite disparate forms of Veblen's bitter social critique, Commons's study of the legal foundations of capitalism and Mitchell's painstaking statistical analysis of business cycles.

In many cases, writers did not go far beyond simply justifying their preoccupation with historical fact and statistics by criticizing the classical school for relying excessively on abstraction, deduction and the assumption of self-interest. They then described their own work as realistic, inductive and cognisant of a variety of ethical standards. Early on many had earned themselves the designation '*Kathedersozialisten*' (socialists of the professorial chair) because of their advocacy of state intervention in the economy and these leanings were later evident also among the American institutionalists. But by no means all spoke from the direction of the political left. Often they merely voiced objections to what they saw as an excessive academic preoccupation with markets and price theory. One spokesman for American institutionalism, referring to the meaning of the term 'the economy', said: 'Institutionalism proposes to find that meaning in the interplay of institutions and technology … just as classical theory has sought the meaning of the economy in the interplay of wants and scarcity.'[1] It may therefore be a fair comment that the institutionalist critique of economics has been fairly amorphous and that in effect its many expressions have had in common only the wish to see what was described earlier as an integration of theory and institutional fact. Since this sentiment had its heyday in the second half of the 19th century, it is perhaps natural that it should have been channelled into the evolutionary, developmental approach which many then considered to be the most respectable scientifically, in contrast to the more analytical bias of the 18th and early 19th centuries, to which there was a return in the 20th century.

This study will have as its *raison d'être* the sentiment which inspired institutionalists in the past, but it will not be based on the institutionalists' most usual approach, that is, the close attention to cultural evolution. Instead, the approach will be wholly analytical. The objective will be to develop a conceptual framework in terms of which it can be shown, first, how a knowledge of institutions fits into our understanding of economic questions and, second, why the presuppositions of micro-economic equilibrium theory make the accommodation of received theory to the institutional peculiarities of specific eras and areas so very difficult. For these purposes it will be necessary to delve into some epistemology. The relation between theory and the particular events of our everyday experience will be considered in some detail. It will then be seen that the question of how institutions may be brought to the closer attention of theorists is a part of the wider question of what we are to understand by a fact in economics. This question is susceptible of two answers and it will be argued that the distinction between these two types of fact is vital to economics. The outcome of all this will differ greatly from the most usual institutionalist analysis. In fact, it will have a greater affinity with the views of the intellectual descendants of the other side in the *Methodenstreit*, who seem also to raise their voices in dissent, but within the community of theorists itself.

1.2 Three aspects of institutions

When economists speak of the institutional framework of an economy they seem to be referring to a set of very diverse entities. An institutional framework may include customs, usages, norms, attitudes and even fashions, a monetary system, a political constitution, a tax system, laws of contract, of inheritance and of land tenure, established means of collective bargaining and many more. It may seem difficult to find common elements in such diversity. However, for the purpose of this study three aspects of institutions are important. It must be stressed that they are important in the present context[2] and not necessarily the most important in any context. Furthermore, since it may not be possible to consider under these three aspects anything anyone has ever called an institution, they may also be taken to delimit the concept of an institution for present purposes.

1.2.1 Consistent conduct

An institution may be described in terms of a possibly unlimited number of statements about consistencies in the conduct of individuals.

Consistent conduct must here be understood as a regular (or fairly regular) conjunction of a type of conduct with a type of situation, that is, an individual's conduct is consistent when he regularly does something, or refrains from doing something, when a certain situation arises. However, this should not be taken to imply that people sometimes behave like automatons, or that institutions are based on conditioned reflexes. Customary or conventional conduct, such as serving refreshments to visitors or making payment by cheque rather than by some other means, is quite compatible with premeditated and purposeful action. Consistent conduct could be rephrased as *conduct appropriate to certain occasions or circumstances*. Using the means–ends terminology, one could then say that certain ends are appropriate to certain occasions in the sense that the occasions present opportunities for doing something agreeable, or call for conduct considered correct, kind, polite, in good taste and so on. When ends are unrelated to occasions, it may nevertheless be appropriate to use conventional means for achieving them, because such means may seem to be the most efficient, most convenient or least risky under the circumstances. In fact, it has been conjectured, as in the conjectural history of money, that the means for achieving some types of ends become institutionalized because they are the most efficient and that thereafter they continue to be the most efficient because they are institutionalized.[3]

Here, the word appropriate has of course been used in different senses, namely, appropriate according to sentiment and logically appropriate. In action, however, the two senses may often become intertwined. For instance, a man may be on his best behaviour not only because he feels that certain occasions call for certain conduct, but also because he wants to make a good impression in order to obtain a more remunerative position in order to give his family a better life and so on. He may therefore consider his conduct appropriate in both senses even though he may know of quicker ways of giving his family a better life. More generally, in calculating whether x or y is the more efficient means of achieving z, one may have to take into account that x is, say, the more socially acceptable or simply the more decent of the two means and that one values being accepted socially or being decent. In other words, the appropriateness according to sentiment may be a factor in calculating costs and logical appropriateness, so that x may be appropriate in both senses.

However consistent conduct may also include behaviour which one would perhaps hesitate to call premeditated and to which the concepts of means and ends would therefore be inapplicable. Customs, like habits, may be followed without thought. For instance, an office worker may have a cup of tea at every break. If this has continued for many years one may not be inclined to say that the worker plans to have it or even wishes to have it. The worker simply does what is appropriate to the occasion. However, uncontrollable behaviour, like flinching one's eye during an eye examination, would clearly not be regarded as institutionalized.

It is not difficult to see that customs, usages and norms may be described in terms of what people are likely or are unlikely to do in certain circumstances. But even, for example, the institution of banking may be described in terms of what, under certain circumstances, is likely to be done by a client with funds coming in and with debts to be settled, by a teller doing his daily work, by a businessman wanting to expand his stocks, by a bank manager considering an application for a loan and so on.

Of course, this would not describe how the appropriate conduct of the various parties fits together, that is, how the banking system functions. In this case the functioning of the whole system may be deduced from the institutionalized behaviour (or appropriate conduct) of the various parties and it is no doubt part of the business of economics to make such deductions. However, it may be argued that the functioning or overall order of a system must be distinguished from what is actually institutionalized, since there are some cases in which the functioning is clearly not part of what is meant by an institution. It may be said, for instance, that the telephone is an institution in developed countries. To understand what is meant by this, one does not have to know how the electronic equipment works, how repairmen locate faults or how some of the funds paid by subscribers find their way into the repairmen's pockets at the end of each month. All that is meant is that a large number of people find it appropriate to reach for a telephone when they want to communicate with someone some distance away.

It could be argued in an analogous way that the institutional part of market forms also comes down ultimately to ideas on appropriate conduct. In the case of a produce market which in some respects comes close to the model of perfect competition, the institutional basis is surely not the mere fact that there are many farmers, places where buyers and sellers can meet and so on. Rather, it is that farmers want, or think it fit and proper, or simply regard it the natural state of affairs, to farm individually rather than to combine into corporations or communes, and to compete against each other rather than to club together in selling. Of course, technical agricultural factors may be involved as well, in which case the market does not have a purely institutional basis.

Finally, institutions may also be described in terms of what is inappropriate conduct or what people are unlikely to do. In describing the political systems of Britain and the United States, for instance, it may be worth noting that a British or American general is unlikely to attempt to seize the government or, if he did contemplate such an action, that he is unlikely to find enough people to support him. In an analogous way one can say that the institution of private property entails that a buyer not only expects to be able to take physical possession of what he buys, but also expects others to (consider it appropriate to) refrain from using his purchase, even though they may desire it and it may be physically feasible for them to use it. If his expectations should be disappointed, then (on the positive side again) he would expect the police to regard it appropriate to act on his complaint.[4]

1.2.2 Consistent conduct as a fragment of the whole

The set of institutionalized or consistent conduct does not embrace all conduct. While an institution always implies consistent conduct, it also has the connotation that there is conduct which is otherwise.

It would make no sense to speak of institutionalized conduct in a world in which all conduct is institutionalized. If there are conventional ways of doing things, there must be also other ways of doing things. If there are customary greetings, gifts and fringe benefits, there must also be other greetings, gifts and fringe benefits. While it is possible to imagine a primitive society in which all conduct is determined by biological needs and ritual, the ritual would be called institutional only in contrast to the situation in more familiar societies. In other words, an institutional framework is not a comprehensive system. Institutionalized or consistent conduct must always be seen as conjoined with action that can be described as unusual, novel, creative, unique and so on. All this may be quite obvious, but the apparently partial consistency of conduct, as will be shown later, is something which economic theory has found very hard to handle.

1.2.3 Consistent conduct as empirical orientation

A knowledge of institutionalized or consistent conduct may be used as a means of empirical orientation, that is, institutions may be 'points of orientation'.[5]

In the ordinary business of life, people use a knowledge of consistent conduct as an aid in interpreting the actions of others, that is, in making sense of the doings of their fellow men, or as a guide to action in planning their own action. When a knowledge of consistent conduct is acquired by induction for these purposes, it consists of empirical facts which are quite independent of theories about the function of institutions in a social order and of the question of why institutions are, or how they came to be, what they are.

1.3 On the empirical content of economics

It is a fair question whether economics can have an empirical content, or if it can, whether it has, or if it has whether the logical form of its empirical content is fully understood. Since institutions, viewed as empirical facts, are obvious candidates for filling any possible vacancies in this empirical content, the issues that may be raised in seeking answers to this question will form much of the subject matter of the following chapters. The preliminary remarks that will be made here will try to show, first, that the answers are at the very least not immediately apparent, second, that economics has to contend with particular difficulties in regard to knowledge and, third, that micro-economic equilibrium theory more or less ignores these difficulties.

1.3.1 Theory without empirical content

For much of its history the main body of economic theory was associated with the advocacy of free trade and of an unrestricted market mechanism. It is not immediately apparent whether there is an empirical aspect to an argument urging that certain measures be taken for the greater happiness of mankind. Certainly, in the hands of Adam Smith the analysis of a natural (and by implication a desirable) order was based on a keen study of institutions. We have the opinion of one of his students that the fourth and last part of his lectures on Moral Philosophy, which 'contained the substance of the work he afterwards published under the title of … the *Wealth of Nations*', considered 'the political institutions relating to commerce, to finances, to ecclesiastical and military establishments'.[6] In subsequent and more formal developments of theory the notion of natural order usually had some role to play, but its institutional basis receded and was almost gone by the time of the neo-classical equilibrium model. Enthusiasm for promoting the market economy seems to have waned as well; the notion of equilibrium is not necessarily associated with a desirable order and its debt to the reformist zeal of former times is largely overlooked. The result is an equilibrium concept which seems peculiarly unrelated to the particulars of actual situations or at least to any that anyone actually claims to know.

One may wish to see economics as the pure logic of choice. In that case, economics cannot have an empirical content. 'Like logic and mathematics', said von Mises, economics 'is in us; it does not come from without'; 'no experience, however rich' could disclose it; it is derived from a 'logical analysis of our inherent knowledge of the category of action'. Furthermore, in von Mises's opinion, there 'are no such things as a historical method of economics or a discipline of institutional economics'.[7] Irrespective of whether one agrees with von Mises, one must admit that the logic of choice is valuable in handling factual material. Even an applied economist has said that in so far as economic theory is useful 'in helping us to take decisions on policy, it is the simple, most elementary and in some ways most obvious propositions that matter' and he mentioned, among other things, opportunity cost and the maxim that bygones are bygones.[8] However, such usefulness presupposes that we know something to which the logic can be applied. It is said that the logic of choice needs special or subsidiary assumptions, some empirical input on which it can work and in this regard we are apparently left to our own devices. The only guidance that von Mises offered was that the 'question whether or not the real conditions of the external world correspond to these assumptions is to be answered by experience'.[9]

One may of course suppose that statistical time series are the empirical raw materials of economics and provide economics with an empirical content. How the logic of choice may be applied to statistics is again not immediately apparent but in the case of equilibrium models there are at least some ideas, for the work of some econometricians seems to be based on the belief that statistics can be incorporated into equilibrium models. However, this raises another question. Are statistics themselves the empirical facts in which economists are interested, or does interest really lie in certain relations that are meant to be distilled from statistics? The methods of econometrics seem to indicate the latter interest. However, the distillation process has not proved to be an easy matter. Professor Hutchison has recently assembled a number of quotations in which eminent economists express their dissatisfaction with the position of empirical fact in their subject. Among them is the comment by Professor Leontief that: 'In no other field of empirical enquiry has so massive and sophisticated a statistical machinery been used with such indifferent results.'[10]

Also in the factual field are studies, often with merely a superficial relation to theory and almost disparagingly called descriptive, which give an account of, say, the marketing of an agricultural product, the protection of an industry or the progress of an anti-inflationary policy and so on. Here also one may ask whether the various episodes of such reports constitute the facts in which we are interested or whether there are lessons to be learnt (or distilled) from history. The point at issue is the same as that raised in the case of statistics, because statistical time series also record aspects of the events of the past.

1.3.2 Empirical constraints versus empirical past

There is something in the nature of a social science which makes the question of an empirical content far more complicated than it is in the natural sciences. In the latter there is a straightforward relation between knower and known, whereas in economics, as in all the social sciences, the knower's known includes other knowers whose known includes other knowers and so on ad infinitum. The matter may be stated differently. 'Economics is a study of mankind in the ordinary business of life', as Marshall said, and the ordinary business of life consists largely of a conscious effort to achieve certain ends. It therefore involves decisions on appropriate means and these must be based on knowledge or what is believed to be knowledge. Knowledge of economic action is therefore knowledge partly of practical knowledge, and this practical knowledge within knowledge carries with it peculiar difficulties which must be considered more closely.

The individual who is the subject of economic enquiry has to decide which are the most efficient means for attaining a desired end under the conditions he expects to prevail between the present and the time of his expected success. However, by most conceptions of what is economic, his actions will necessarily involve other people, so that both the expected conditions and the efficiency of the means depend on what others will do. In order to plan with precision, the economic subject has to be able (at the least) to keep his options open until he knows what others are doing. But the others are in the same position. They also cannot calculate their best course of action until everyone else has committed themself to their course of action. Because of the interrelation of knowers and known, attempts at fully informed action would require a grand pre-reconciliation of all action or lead to a waiting game in which no one would ever get started. But there is no computerized *tâtonnement* and no deadlock; people do act and we believe they act purposefully.

The problem of how purposeful action is possible in circumstances such as these has been part of the underlying theme of Professor Shackle's many writings. He has referred to it as 'the epistemic problem, the problem of how the necessary knowledge on which reason can base itself is to be gained, the problem of what to suppose that men will do when time's sudden mockery reveals their supposed knowledge to be hollow'.[11] The solution he has put forward is that there is something between a completely calculable world and a 'cosmos in which no act places any constraint whatever upon the character of the sequel', that there is 'bounded uncertainty', that 'there are constraints as to what range of diverse things can happen'.[12] His vision is of a 'kaleidic society' in which now one and then another rival orientation and rival interpretation gains the ascendancy and the arrival of 'the news' can change the whole picture.[13] Nevertheless, a bounded uncertainty needs constraints, even if they are not universally agreed upon. Though Shackle does not say so himself and though it is not really in the spirit of his work, institutions viewed as consistent conduct in the midst of novel action can augment just such constraints on uncertainty.

Shackle acknowledges the influence of Hayek's study of the role of knowledge in economic affairs. However, Hayek's solution to the epistemic problem, characteristically directed towards an understanding of the overall order of a market economy, is somewhat different. He focuses attention upon knowledge of a different kind of fact. For the businessman planning a new venture there is 'hardly anything that happens anywhere in the world that might not have an effect on the decision he ought to make'. But he does not have to know what everybody else is doing, for the great significance of the price system is 'how little the individual participants need to know in order to be able to take the right action'. He merely has to watch various prices because all that is relevant to him in the doings of others is reflected in them. The market is therefore 'a mechanism for communicating information', but the information it conveys is not about constraints on the range of things that *can* happen, but rather about 'particular circumstances of time and place' or 'of the fleeting moment', that is, about events that *have* happened or *are* happening.[14] The distinction refers once more to the two sides of empirical knowledge which were alluded to earlier with regard to statistics and descriptive studies and which will be considered in some detail in Chapter 2.

1.3.3 Observer perspective versus operating subject perspective

Because of the circumstance that the knower's known includes other knowers, there is a tendency in the social sciences to treat of two kinds of knowledge, namely, that of the observing social scientist and that of the people he studies. Often the difference between them is that the observer, uninterested in the details of his subjects' affairs, deals only with the broad categories of his subjects' knowledge, such as (in economics) tastes, production coefficients, factors of production and so on. In the standard neo-classical equilibrium model this tendency is very marked. The equilibrium theorist is not concerned with the details of how his subjects go about their daily lives, but rather with the general features of the order to which their actions give rise. Unlike the two writers considered in section 1.3.2 (Shackle and Hayek), the equilibrium theorist is unconcerned with the details of his subjects' affairs to the extent that he does not pay much attention to their knowledge problem, but concentrates on the order that would prevail if his subjects did not have such a problem.

Hayek pointed out long ago that there is often a confusion in economics about the concept of a datum. 'Datum means, of course, something given, but the question which is left open, and which in the social sciences is capable of two different answers, is to *whom* the facts are supposed to be given ... to the observing economist or to the persons whose actions he wants to explain'. If it is to the latter, the market can have an information function only if the facts are not given equally to all. This was Hayek's interest in the matter.[15] However, with regard to the knowledge of the observer and the observed, the important question is not to whom but what is given. When a problem handled by equilibrium theory is qualified by the expression 'given tastes', this expression signifies much less in the economist's knowledge than it supposedly does in the knowledge of his subjects. Hayek was himself to write later (quoting Pareto) that it would be absurd to think that in the case of such entities as tastes the economist is able to 'fill in all the blanks', that is, to ascertain and specify tastes in detail.[16] In other words, the only significance of 'given tastes' for the economist's knowledge is the fact that they *are* given, that is, that they do not change during the course of the analysis, and that they have the properties of all tastes, namely, that they are logically consistent and obey the rule expressed as the diminishing marginal rate of substitution.

Since these elements of the economist's knowledge, and others such as diminishing returns, are comparatively simple, they lend themselves to treatment as axioms in purely logical or axiomatic constructs and equilibrium is the non-specific solution to a mathematical optimization problem. However, the axiomatic constructs are meant to represent something empirically real, albeit in a very idealized form. Equilibrium is therefore visualized as a balance of interests, a harmonious reconciliation of the purposeful actions of many individuals. In this way the economist has been able to use the logic of choice to arrive at a rigorous analysis of a natural order. In view of the importance of the notion of natural order in the history of the subject, this is a considerable achievement.

But it can be attained only at the expense of putting a great burden on the supposed knowledge of the subjects of the enquiry. Even though the topics of false trading and of information costs have been paid some attention and such theoretical devices as price criers, *tâtonnement* and re-contracting have been considered, the question of how the economic subjects actually manage to bring about a harmonious reconciliation is left largely unanswered. The most usual procedure is to side-step the issue by assuming that the subjects' knowledge is complete and that there is therefore perfect foresight. As

Shackle put it, the epistemic problem has been ignored, and in Hayek's view, the function of the price system has been pre-empted.[17]

The empirical part of the economist's knowledge is therefore far more modest than that which he attributes to his subjects' knowledge. To the economist the expression 'given tastes' signifies a few relatively simple propositions. To the subjects he studies it means knowing each other's orders of preference for a vast and diverse number of goods and services, the ratios at which these goods and services are substituted for each other when a person already has a combination of them at his command and so on. In other words, the economist can only achieve his neat analysis by assuming that the subjects of his study have full information on something about which he knows next to nothing, and of which the little he does know is of questionable empirical validity.

1.4 Purview of the following chapters

The discussion in sections 1.2 and 1.3 may now be used to give more definition to the tasks set for this study in section 1.1. In the turn of phrase adopted in this chapter, one of the problems to be investigated is why equilibrium theorists appear to have little use for a specific knowledge of conduct appropriate to occasions and circumstances.

It may occur to one, in considering this question, that there seems to be very little in an equilibrium model that can properly be called conduct of any kind. The range of interest of value-theoretic, neo-classical equilibrium models is such that people are not required to execute shrewd or imaginative plans; all they have to do is produce and buy. However, this in itself implies some institutionalized conduct. People in equilibrium models are accustomed to trade and have an impeccable respect for property rights and the law of contract. There never is anybody in an equilibrium model who decides that the best way of obtaining a desired object is to hit its present owner over the head. Some institutionalized conduct is therefore presupposed when individuals are cast in the roles of producers and buyers. Furthermore, each individual calculates very precisely what he should do in his capacity as a producer and as a buyer. If one can refer to this as his appropriate conduct, one finds that such conduct is based ultimately on his own and every other individual's preference field, the state of technical knowledge and the availability and ownership of resources, about which he as everyone else (except the observing economist) is fully informed.

It seems therefore that equilibrium theorists feel no need for any information on what people are likely to do on some occasions and in some circumstances because they assume that *all* of their subjects' conduct is already fully knowable from information on certain apparently very basic factors. In the context of an equilibrium model, occasions and circumstances can come about, in any case, only by the confluence of the conduct of various individuals, since exogenous factors such as the weather, earthquakes, epidemics and the like are not unnaturally left out of consideration. There is then quite obviously no point in seeking a knowledge of the conjunction of types of conduct with types of circumstances, because all of that is already taken care of by those basic factors which the equilibrium theorist indeed does not claim to be able to ascertain specifically, but which he has identified non-specifically.

In view of the dominance of the notion of equilibrium in economic theory, the all-embracing nature or comprehensiveness of an equilibrium system built on preference fields, technical knowledge and the availability and ownership of resources will therefore stand in the way of any attempt to let institutions play a more substantial part in economic analysis. It will be argued in a later chapter that, of these four factors, it is the preference field idea to which the comprehensiveness of equilibrium theory can really be attributed. If this is so and one feels, as many have in the past, that economic theory can be criticized for neglecting institutions, one must try to show that there is something wrong with the preference field idea. Such an attempt will be made in this study.

However, its primary purpose will not be a critique of neo-classical economics, but rather the development of an alternative conceptual framework which could ultimately enable the economic theorist to use institutions as a means of empirical orientation. (The present study can of course go only a short way along this route.) The new conceptual framework will be evolved, in the course of an epistemological investigation of micro-economic theory, from the question of the acquisition of knowledge or from what Shackle calls the epistemic problem. In the process, economic concepts will be judged not only by their capacity for creating a formal logical order but also by their capacity for making known the empirically real. From this point of view, the idea of preference fields could be criticized in many ways. One could say, for instance, that it is not practically feasible that they should be known, or that they do not exist and therefore cannot be known, or simply that as theoretical devices they are heuristically unjustifiable. In the approach that will be adopted here it will be seen that these apparently incompatible objections do not in effect differ very much.

The approach is to ignore the distinction between the knowledge of the economist and the knowledge of the people studied, to deny that the economist's capability for acquiring knowledge is in any way different from that of the person who executes economic plans. The observer and the observed will be on the same footing. The only distance between them is to view the observer as merely one potential knower among many who try to learn and to make sense of the world around them.

Then, if it is absurd to suppose that the economist can ascertain people's preferences in detail, it is also absurd to suppose that anybody else can; and if it is possible for the person in the street to use a knowledge of institutionalized conduct as a point of orientation, it is also possible for the economist to know such institutionalized conduct and to use it as a point of orientation. In other words, it is then possible for institutions to play a part in the empirical content of economics.

1.5 Plan of the following chapters

Chapter 2 outlines and explains the essentials of the conceptual framework which is being proposed. Chapter 3 deals with the deterministic presupposition which, as will be argued later, has partly guided equilibrium theorists. It also considers certain objections to and qualifications of determinism in economics and then compares these and determinism to the conceptual framework outlined in the previous chapter. The chapter puts forward the proposition that micro-economic theory is largely a blend of axiomatic constructs and deterministic models. It then tries to isolate the deterministic elements. Chapter 4 considers the role which the notion of rational action plays in what was dealt with in the previous chapters. Chapter 5 briefly traces the development of common-sense notions of needs, tastes and preferences into the concept of an ordinal preference field, and tries to show that it is extremely unlikely that an ordinal preference field, or whatever it represents, can be both consistent over time and comprehensive (in the sense that it is behind all the choices of an individual) and that the concept, as it has evolved in the 20th century, is likely to be based on a confusion between different logical forms of empirical fact. Chapter 6 discusses the nature of an economics which does not rely on the concept of a preference field, but which provides for empirical orientation by means of institutions.

Notes

[1]   C.E. Ayres, 'The Co-ordinates of Institutionalism', Papers and Proceedings of the American Economic Association, in *American Economic Review*, May 1951, p 52.

[2]   A more usual approach is to look for common principles in the evolution of institutions. See, for example, F.A. Hayek, 'Notes on the Evolution of Systems of Rules of Conduct' in *Studies in Philosophy, Politics and Economics*, (London: Routledge, 1967) pp 66–81, in which he investigates the interplay between rules of conduct and the 'overall order' of an economy or a society. Some sociological theories stress the complementarity of the functions of institutions, which sometimes leads to an analogy between societies and biological organisms. This is not the approach adopted here.

[3]   Carl Menger elaborated this view considerably in his exposition of money. It first appeared in the chapter on money in his *Grundsätze der Volkswirthschaftslehre* of 1871. In his *Untersuchungen über die Methode der Socialwissenschaften* of 1883 he extended the argument to customs, common law, languages, towns, and so on, and argued that the logical form

of the analysis is the same as that of price theory. (This idea will be explained in later chapters.) When he repeated the argument about the origin of money in *Geld* of 1909, he likened the difference between money and commodities to the difference between roads and other pieces of ground. See *The Collected Works of Carl Menger* (London: London School of Economics Reprint, 1933–6) vol 1, pp 250–60, vol 2, pp 172–83 and vol 4, pp 3–27.

4   The distinction between physical possession, which even a receiver of stolen goods gets, and legal ownership is a dominant theme in John R. Commons, *Institutional Economics* (Madison: University of Wisconsin Press, 1961; first published New York: Macmillan, 1934). Commons, generally recognized as having been one of the three leading figures in American Institutionalism, maintained that orthodox economics did not make the distinction and that this went a long way towards explaining why institutions played such a faint role in economic theory. He distinguished between exchange and transactions. 'Transactions ... are not the "exchange of commodities" in the physical sense of "delivery", they are the alienation and acquisition, between individuals, of the rights of future ownership of physical things as determined by the collective working rules of society' (p 58). In orthodox theory, he maintained, exchange and transactions were identical and this gave a double meaning to wealth, namely, 'the physical meaning of holding the materials of nature for one's own use in production and consumption, and the proprietary meaning ... namely the right to exclude others and to withhold from them what they want but do not own' (p 302). By neglecting the proprietary meaning, orthodox economists 'concealed the field of institutional economics' and it was 'this concealed ownership side of the double meaning of Wealth that angered the heterodox economists' among whom was Marx' (p 55). By assuming physical possession and legal ownership to be identical, orthodox economists could ignore statute law, ethics, customs and judicial decisions when constructing 'a theory of pure economics based solely on the physical exchange of materials and services' (p 56).

5   This is an expression used by L.M. Lachmann in *The Legacy of Max Weber* (London: Heinemann, 1970) p 38. In the place quoted he does not restrict the term to institutionalized conduct. However, later in the book in an essay entitled 'On Institutions', he says that institutions provide 'means of orientation' (p 49), that the rules of a game 'constitute a set of orientation points' (p 61) and, quoting and translating Weber, that institutional norms are used by a certain group as 'a means of orientation of their (legal or illegal) acts because certain expectations concerning the conduct of others attach to them' (p 62).

6   The student was a Mr Millar who later became Professor of Law at Glasgow University and apparently a close friend of Adam Smith. The extracts appear in a long passage quoted in and probably solicited for Dugald Stewart's *Account of the Life and Writings of Adam Smith, LLD*. The source here is a reprint of the short biography included in Smith's *Theory of Moral Sentiments* (London: Bell, 1892) p xvii. See also Nathan Rosenberg, 'Some Institutional Aspects of the Wealth of Nations', *Journal of Political Economy*, 68, 1960, pp 557–70.

7   L. von Mises, *Human Action* (London: Hodge, 1949) pp 64 and 66. The term 'pure logic of choice' seems to have been coined by Hayek and von Mises does not use it, having himself made up the term 'praxeology'. The two terms have more or less the same meaning except that Hayek's has a slightly pejorative connotation, since he does not agree with von Mises's views on the purely a priori nature of economics.

8   Ely Devons, 'Applied Economics: The Application of What?' in *The Logic of Personal Knowledge, Essays Presented to Michael Polanyi on his Seventieth Birthday* (London: Routledge, 1961) pp 155–69.

9   L. von Mises, *The Ultimate Foundations of Economic Science* (Princeton: van Nostrand, 1962) p 44.

10. T.W. Hutchison, '"Crisis" in the Seventies: The Crisis of Abstraction' in *Knowledge and Ignorance in Economics* (Oxford: Blackwell, 1977) pp 62–97. The quote in the text is on p 71. The original source is the Presidential Address to the American Economic Association, 1970, in *American Economic Review*. Others quoted are Professors Ragnar Frisch and Harry Johnson, Lord Kaldor, Sir Henry Phelps Brown and Mr G.D.N. Worswick, Director of the National Institute for Social and Economic Research. The last-named has been especially outspoken: Those responsible for 'some econometric theory' are not 'engaged in forging tools to arrange and measure actual facts so much as making a marvelous array of pretend-tools which would perform wonders if ever a set of facts should turn up in the right form'. 'There now exist whole branches of abstract economic theory which have no links with concrete facts and are almost indistinguishable from pure mathematics.' The only 'distinguishing feature is that some of the axioms and some of the terminology show traces of the ancestry of this particular branch of mathematics, which originated in the distant past in some real economic question'. Then there is Sir Henry's complaint that 'the human propensities and reactions' which economics 'purports to abstract are not in fact abstracted ... but are simply assumed'.

11. G.L.S. Shackle, *Epistemics and Economics* (Cambridge: Cambridge University Press, 1972) p 447.

12. Shackle analyses the idea of bounded uncertainty in great detail in *Decision Order and Time in Human Affairs* (Cambridge: Cambridge University Press, 1961). The quotations are from pp 4 and 271.

13. For something of the flavour of the kaleidic society, see Shackle, *op. cit.* (note 11) pp 76–9.

14. F.A. Hayek, 'The Use of Knowledge in Society' in *Individualism and Economic Order* (London: Routledge, 1949) pp 80, 81, 84 and 86. The same idea could also have been gleaned from the articles 'Economics and Knowledge' and 'The Meaning of Competition' in the same book. In all three articles Hayek of course says a great deal more than is reported in the text.

15. Hayek, 'Economics and Knowledge', *op. cit.* (note 14) p 39. His italics.

16. Hayek, 'The Theory of Complex Phenomena' in *Studies in Philosophy, Politics and Economics*, *op. cit.* (note 2) p 35.

17. Shackle, *op. cit.* (note 11) pp 221 and 447. Hayek, 'The Meaning of Competition', *op. cit.* (note 14) pp 94f.

2

Ex Post and *Ex Ante* Facts

2.1 A brief indication of the distinction

In this chapter I shall deal with an epistemological distinction which will form the basis of the later analysis. In order to make it easier for the reader to grasp this distinction, I shall begin with an illustration drawn from familiar material.

Let us consider what rising price indices tell us about inflation. Do they tell us *why* prices *are* rising or do they tell us *that* prices *have* risen? Do they convey the sort of information we would need to devise a policy to counter inflation, that is, do they provide us with a guide to action in the future? Perhaps in some cases they would, but it is far more likely that we should have to regard them (with due allowance for the index number problem) merely as a record of the course of inflation over a particular period in a particular area. It hardly seems likely that a mere scanning of statistical time series of prices would bring to mind some kind of laws governing price increases which would enable us to control price changes in the future. Common sense would tell us to distinguish between facts that merely record past events and facts that somehow relate to the structure of the economy. If we want to recommend an anti-inflationary policy we may well look at price indices to see what actually has taken place, but we would know that this in itself would not be enough. We also would need knowledge of certain structural features of the economy. We would turn, according to our inclinations, to the quantity theory of money, the inflationary gap analysis, the degree of price competition in markets, especially in labour markets, and so on.

Let us suppose that we had not only price indices but also other information ancillary to them. Let us suppose that we had a copious description of the circumstances and intentions of each and every seller at the time when he changed his price and of each and every buyer at the time when he offered to pay a higher price. (The practical difficulty of gathering such information is of no importance here.) We should now be able to satisfy ourselves that we have a detailed knowledge of how it was that a particular bout of inflation came about. However, even all this information would not necessarily tell us anything about what we could do to prevent inflation in the future. If we were very lucky we might notice certain common elements in the situations in which prices were raised and so have a clue to some hitherto unknown structural feature of the economy. However, the history of economics shows that the record of such things simply springing to mind is very poor indeed. It is far more likely that we should notice, on looking back over the course of inflation, so many events that seem to be of a purely coincidental and fortuitous nature that we should not expect them to recur, and so we would be left without a guide to possible action.

I have chosen this rather extensive illustration because I hope that what I have said will seem obvious to the reader. The point was to show that it is possible to distinguish between two entirely different orders of facts. I shall try to explain the distinction at some length in a later section. To facilitate the discussion in the meantime, I shall provisionally label '*ex post*' those facts which owe the meaning they have for us to their position in a possibly unique and fortuitous or stochastic course of events. The other order of facts, which relates to structures, that is, to not altogether transient patterns, and which may, therefore serve as a guide to our purposeful action, I shall label '*ex ante*'. Why these terms, which are, of course, in current use in economics, seemed to me appropriate will, I hope, become apparent as I proceed.

A few words on the purpose of distinguishing between *ex post* and *ex ante* facts may here be in place. The distinction is by no means always as easy to perceive as in the illustration provided. In fact, I shall attempt to show that economists frequently mistake *ex post* facts for *ex ante* facts; so much so that I want to suggest that not very many serious attempts are made to isolate true *ex ante* facts, at least in a useful form. If this could indeed be established, it would follow that economists cannot have much to contribute when others turn to them for advice on action to be taken. I believe one could show, for instance, that even the inflationary gap analysis, cited earlier as apparently concerned with an occasional structural feature of the economy, really dresses up *ex post* facts as *ex ante* facts, and indeed only has meaning in an *ex post* context. In this work, however, I shall confine myself to trying to show how a confusion between *ex post* and *ex ante* facts has manifested itself in micro-economics.

2.2 Plan of the chapter

If one were to ask a number of economists whether they believed in a universal determinism, the majority would perhaps take up an agnostic position. In any case, whatever their opinions, they are likely to feel that they can carry on their work as economists quite well without deciding the issue. One should not, however, underestimate the surreptitious and pervading influence on academic work of subconsciously held philosophical presuppositions. I want to suggest in sections 2.3 to 2.7 that a widely held view of the vocation of science is largely responsible for, and makes entirely understandable, the frequent failure to distinguish between what I have called *ex post* and *ex ante* facts. In sections 2.5 and 2.6, I shall also deal with two types of objections, raised by economists, against the application to economics or to the social sciences in general of this metaphysical view of what are the proper aspirations of science. I shall do all this in order to prepare the ground for an outline of a conceptual framework based on the distinction I have described. I shall give this outline in section 2.8. In section 2.9 I shall discuss the selection and testing of hypotheses in so far as this pertains to the distinction here made. Finally, in section 2.10, I shall make a few brief remarks on the potential role of a philosophy of science applied to economics.

2.3 The deterministic presupposition

The spirit of a wholly deterministic outlook, and of the corresponding scientific programme, was strikingly captured in a famous passage by Laplace, the great astronomer and mathematician:

> Given for one instant an intelligence which could comprehend all the forces by which nature is animated and the respective situation of the beings who compose it – an intelligence sufficiently vast to submit these data to analysis – it would embrace in the same formula the movements of the greatest bodies of the universe and those of the lightest atom; for it, nothing would be uncertain and the future, as the past, would be present to its eyes. The human mind offers, in the perfection which it has been able to give to astronomy, a feeble idea of this intelligence. Its discoveries in mechanics and geometry, added to that of universal gravity, have enabled it to comprehend in the same analytical expressions the past and future states of the system of the world. Applying the same method to some other objects of its knowledge, it has succeeded in referring to general laws, observed phenomena and in foreseeing those which given circumstances ought to produce. All these efforts in the search for truth tend to lead it back continually to the vast intelligence which we have just mentioned, but from which it will always remain infinitely removed.[1]

Laplace went on to argue that the ever-present gap between the human and the 'vast' intelligence makes a theory of probability essential. However, the ideal that science should ever aspire to this vast intelligence is still deeply rooted in our thinking. Given the governing laws and the initial state, or in mathematical language, given the relevant set of differential equations and the boundary conditions, we can trace the evolution of any system. The genesis of this view perhaps owes something to man's early scientific preoccupation with astronomy. In its own special sphere, astronomy has indeed come close to that vast intelligence, as Laplace pointed out. Can we not go to a planetarium and ask to be shown a representation of the night sky as seen from any point on the surface of the earth and at any date, in the past or in the future, that we care to name? And where else in science is this kind of thing possible? It was natural perhaps, in the excitement of the profound scientific discoveries of the 17th and 18th centuries, that people should have expected the same progress to be possible in other spheres. 'The regularity which astronomy shows us in the movements of the comets doubtless exists also in all phenomena', exclaimed Laplace.[2] However, astronomy and the related classical mechanics have proved to be special cases. Here, man was somehow able to discern structures which, in the form of theoretical models, could be thought of as viewed from no particular place or time, so that man in turn could be thought of as an entirely passive and unimportant observer whose actual constraints of place and time did not really matter.[3]

It is well known that in time more and more physicists became sceptical of the deterministic belief, and that some rejected it altogether. Leading physicists have tended to be far more epistemologically aware than, for instance, the leading economists, and some well-informed discussion of the issue came into print.[4] In 1938 we find Sir Arthur Eddington saying:

> Following Laplace, it is assumed that from the complete state of the universe at any one instant the complete state at any other instant, past or future, is calculable. The fundamental laws of nature are then defined to be the laws which, taken all together, furnish a sufficient set of rules for the calculation. To complete our knowledge of the universe we must know, besides the rules, the initial data to which they are to be applied. These data are the special facts. ... But this mode of distinction is possible only in a deterministic universe. In the current indeterministic system of physics there is no corresponding demarcation between the laws and the special facts of nature. The present system of fundamental laws does not furnish a complete set of rules for the calculation of the future. It is not even part of such a set, for it is concerned only with the calculation of probabilities.[5]

Somewhat further on in the book he considers the 'suggestion that a proper reformulation of our elementary concepts would banish the present indeterminism from the system of physics'. His reply is significant in the context of *ex post* and *ex ante* facts:

But the suggestion overlooks the essential feature of the indeterminism of the present system of physics, namely that the quantities which it can predict only with uncertainty are quantities which, *when the time comes*, we shall be able to observe with high precision. The fault is therefore not in our having chosen concepts inappropriate to observational knowledge.[6]

I do not know whether all present-day physicists would agree with Eddington's assertions. He was known for his belief in the *a priori* character of physical laws, and this does not find general favour. Perhaps there are still differences of opinion. The point is that for the most part it does not really matter in the physical sciences. Scientists other than astronomers (and other than economists, one could add) hardly ever have to see themselves as passive observers and predictors of the movements of whole systems. The ordinary scientists in physics and chemistry laboratories, and the men who develop the knowledge that is applied in technology, are engaged in a much more humble task. Prediction in the physical sciences, and the knowledge needed for it, are in the most usual cases conditional on human agency. The sort of knowledge that most scientists are concerned with can be characterized as follows: *If one performs the operation x* in a situation which contains the elements y, then there will appear, possibly after a number of distinguishable intermediate stages, a new situation which contains the elements z, or contains, according to calculable probabilities, either z' or z'', z''', and so on. He is concerned, as we shall see, with one of the two types of what I have called *ex ante* facts – facts which in principle can be used as guides to action.

2.4 The deterministic presupposition in economics

The position is different in economics. By tradition, economists very often concern themselves, as passive observers, with the movements of whole systems, if for 'systems' one reads 'economies'. In the case of the crude extrapolations of some business economists, the analogy to the methods of the old astronomy is obvious. No doubt one cannot accuse the majority of economists of such naivety. However, even in far more sophisticated work there is very often the implication that we are placing ourselves in the position of passive observers of a deterministic system. There appears to be such an implication, for instance, whenever we speak of a determinate equilibrium solution. We may say (1) that we are here using the word 'determinate' in the mathematical sense, (2) that usually we are interested only in the implications of certain parameters and not in tracing a course of events from some specified initial state, and (3) that the models we use for this purpose are not meant to represent, but only to approximate very loosely, not all, but only certain aspects of an actual economy. Nevertheless, such qualifications do not remove the deterministic implication.[7]

We have seen that determinism consists not merely of the notion that we may find regularities by empirical means – *ex ante* facts in my terminology – but rather of the further notion that these facts, which we gather piecemeal, may be built up according to Laplace's vision into a vast model that could explain and predict the entire course of events, given only the state of things at any instant of time. Within the confines of what was probably its original context, namely, classical mechanics, this notion seemed quite well-founded. Problems arise, however, when the mechanical analogy is taken into other fields. In economics it has to be admitted that even so basic a question as how a person will act in a given environment, or react to given stimuli, cannot be answered with anything like the exactness of mechanics, if indeed it can be answered at all. Explanations of the failure of the mechanical analogy in economics have usually followed either one of two themes, namely, that the human will is undetermined or that the great complexity of social phenomena makes a strictly deterministic scientific programme unworkable.[8]

2.5 The free-will argument

Determinism does not fit in with what we think we know of the voluntary nature of our own actions. Those who argue in terms of free will replace the belief in a universal determinism by a dichotomy of a determinate 'dead' nature and a purposefully acting, freely choosing human spirit. (Animals, curiously, are often explicitly excluded from this status.) The long chains of causal reactions stretching through time, which are necessarily a part of the deterministic outlook, are seen as broken whenever they come into contact with a freely acting human being. Often the matter is put in teleological terms. While physical events are determined causally by antecedent events, human action is determined teleologically by imagined events in the future and imagined events do not necessarily materialize. It is not denied that man is influenced by his environment, but his reactions depend on an independent and free will, on his intentions or plans and on his expectations of future events, which in turn depend on his necessarily limited knowledge at the time he acts. No two human beings can therefore be expected to react identically in identical physical conditions.

The dichotomy of determinate nature and free will is probably as common as the presupposition of universal determinism, and as subconscious as well. In many cases one individual probably holds both beliefs, calling forth one or the other to particular purposes. In recent years the most forceful exponent of this position of a conscious rejection of universal determinism and insistence on the dichotomy has been G.L.S. Shackle.[9] He made his point eloquently in a 1974 address:

> If men's thoughts are implicit in their experience, choice is a mere stage in nature's process, an event engendered determinately by other events and serving as a passive link in the course of history. If so, history is not made by men, but merely suffered by them. ... But if choice can arrange the given building-blocks in designs of its own; if thought can manipulate, ex nihilo in some degree, the suggestions offered by the sensations which feed it; if thought can be original, in some true sense; then history can be continuous novelty, not merely in the sense that we have not found the code and secret theme which could tell the whole detailed story from the beginning to the end of time, but because that story does not, at each present moment, exist beyond that moment.[10]

The allusions to Laplace's scheme are here obvious. It should be noted that Shackle does not deny determinism in physical events (nature's process) and that he must suspect his audience of extending this view also to human affairs. Otherwise, what would be the point of his message?

The dichotomy of nature and free will also leads to the view that economics as one of the sciences of human action requires an approach different from that of the physical sciences. The approach that has often been advocated is teleological and hermeneutic. Economic as well as social phenomena in general must be seen as either the intended or the unintended consequences of the purposeful conduct of human beings. Any understanding of such phenomena must then be gained by constant reference to the apparent intentions of acting individuals, or, in the case of unintended consequences of actions, an investigation must at least start with such a reference. Such an approach is sometimes characterized by the term methodological individualism. In so far as it is used to deal with empirical facts, it thus tries to make past events intelligible to us in the ordinary human terms of wishes, beliefs and intentions, and so on. What emerges is not unlike the telling of a story.

An insistence that economic analysis should constantly refer to the meaning which individual actors attach to their actions is a feature of the Austrian approach to economics. In a context in which human action is the subject of an empirical investigation, which is the context with which I shall be largely concerned, the seminal ideas of Menger and von Mises may be associated with the methodological implications of the free-will argument, though it would not be strictly correct to associate them with the free-will argument itself. I shall deal with certain aspects of the approach of the Austrian school of economics in Chapters 4 and 5. Here it may be noted that Menger actually said that economics is unaffected by the question of free will, but he understood pure economics to be a set of what he called laws of thought or, in effect, logical propositions. He seemed to take it for granted that human action is unpredictable, and the application of pure economics to empirical work which he had in mind was not a deterministic one. The main theme running through von Mises's work is the logic of choice (or praxeology) as he called it, and to him the question of free will also appears to have little relevance. His views on free will were rather complicated. In effect he rejected the free-will argument but nevertheless regarded human action to be quite unpredictable because of the complexity of the mind. The conclusion drawn from this 'complexity' argument is often that one should make the best of things by adopting watered-down deterministic methods. Von Mises, however, did not come to this conclusion; in fact he was violently opposed to it. He advocated a strict epistemological division between the physical and the social sciences. Deterministic methods were appropriate in the former but not in the latter. The view that human action is completely unpredictable is to all intents and purposes equivalent to a belief in the freedom of the will.[11]

It is a feature then of the explicit and implicit free-will arguments not only that human action is completely unpredictable but also that deterministic methods are entirely valid in the physical sciences. (In the case of the explicit free-will argument, a juxtaposition of determinate nature and free will seems to be implied by the term free will itself.) Neither assumption is of course essential for the method of making past events intelligible to us by an enquiry into the intentions and beliefs that prompted acting individuals. One may say, and this is in fact sometimes said, that this method is simply interesting in itself and obviously not available to the physical scientist because he deals with things which we do not believe to have intentions and beliefs. This view, however, gives us no *a priori* reason for supposing that the physical-sciences type of search for guides to action could not also bear fruit in the social sciences. Furthermore, one could regard determinism as merely a particular feature of celestial mechanics which also gives us no *a priori* reason for believing that all physical processes may be explained and predicted in the same way. The fact that we can think of innumerable instances in which we could not predict how a person would act would not then be particularly significant because we would then also expect to find many other cases in which we cannot predict what will happen next. It seems to me that this would be a far less constricting outlook. Perhaps the proponents of the free-will argument do not proceed to such an outlook because the case for determinism just seems too plausible or because the fundamental distinction between the social and the physical sciences (which has deep philosophical roots) has to be maintained as a matter of principle.[12] Having seen that the deterministic presupposition introduces difficulties into economics, they reduce its scope, but they are not prepared to abandon it altogether.[13]

2.6 The complexity argument

One often hears the assertion that the phenomena which economics has to contend with are somehow far more complex than the simple phenomena of the physical sciences. Unfortunately, the thought is not followed up very often. I shall refer to an article by Hayek entitled 'The Theory of Complex Phenomena'[14] in which he did follow up the thought and with which, I think, many economists would be in broad agreement. The page numbers cited in the following all refer to this article. On page 34, Hayek sums up the problem succinctly:

> One of the chief results so far achieved by theoretical work in these fields ['the phenomena of mind and society'] seems to me to be the demonstration that here individual events regularly depend on so many concrete circumstances that we shall never in fact be in a position to ascertain them all; and that in consequence not only the ideal of prediction and control must largely remain beyond our reach, but also the hope remain illusory that we can discover by observation regular connections between the individual events. The very insight which theory provides, for example, that almost any event in the course of a man's life may have some effect on almost any of his future actions, makes it impossible that we translate our theoretical knowledge into predictions of specific events.

In other words, the deterministic scientific programme is impracticable for us because our subject matter is too complex. Hayek does not commit himself on the question of determinism. He does not have to, because the question can be decided even within the confines of this presupposition (p 37): 'There may well be valid and more grave philosophical objections to the claim that science can demonstrate a universal determinism; but for all practical purposes the limits created by the impossibility of ascertaining all the particular data required to derive detailed conclusions from our theories are probably much narrower.'

What lesson can be learnt from this? Hayek says (p 35) that 'economic theory is confined to describing kinds of patterns which will appear if certain general conditions are satisfied, but can rarely if ever derive from this knowledge any predictions of specific phenomena'. By way of example he mentions 'those systems of simultaneous equations which since Leon Walras have been widely used to represent the general relations between the prices and the quantities of all commodities bought and sold'. Here it would be absurd to think that we can 'fill in all the blanks'. He continues (p 35):

> The prediction of the formation of this general kind of pattern rests on certain very general factual assumptions (such as that most people engage in trade in order to earn an income, that they prefer a larger income to a smaller one, that they are not prevented from entering whatever trade they wish, etc., – assumptions which determine the scope of the variables but not their particular values).

Not many economists would object to the way Hayek has put this. Difficulties which the deterministic presupposition introduces into economics have been seen and, as in the case of the free-will protagonists, certain adjustments have had to be made to methods and aims.

I should like to take the analysis a little further so that the nature of the complexity can be seen more clearly. By what criterion are we to judge complexity? Hayek says (p 25): 'The minimum number of elements of which an instance of the pattern must consist in order to exhibit all the characteristic attributes of the class of patterns in question appears to provide an unambiguous criterion.' Let us put this a little differently. Let us say that we are interested in an entity y and let us assume – and it is important to remember that this is an assumption – that we can define y in functional form such that $y = f(x_1, x_2, x_3, ..., x_n)$. The degree of complexity of y is now directly proportional to n, that is, to the number of variables that determine it. Here we have taken into account only the 'number of elements' and not their permutations, which could conceivably make y vastly more complex if we thought in terms of a time order, that is, if we were interested, as we often are in economics, in an entity or pattern that 'takes place' rather than 'exists'. (But, I think, the point can be made quite well without these additional complications.)

If y were the orbital motion of a planet or the radioactive decay of a substance (which is a function of time) then one could judge it to be a very simple phenomenon. In comparison, the complexity of the Walrasian general equilibrium system is indeed awe-inspiring. Now, we can say that physical scientists have been lucky to find such simple entities (Hayek even maintains that we classify as 'physical' anything that is simple), and that we may as well face the fact that economists have not. But scientists have not been content just to count on their luck. I suggested at the end of section 2.3 that the sort of fact that a scientist is most usually concerned with is conditional on his own agency. The orbital motion of a planet and the radioactive decay of a substance are among the exceptions to this, for they can be observed without the intervention of the scientist. (This is where the luck lies.) However, suppose a scientist inherits from everyday life a concept y_0 such that it seems that $y_0 = f(x_1, x_2, x_3 \ldots x_n)$ with n fairly large, that is, with y_0 a complex phenomenon. Not knowing what to do with y_0, he says: 'Let me discard the concept y_0 and form another concept y_1 which focuses attention on what happens when I deliberately change x_1.' He now has $y_1 = g(x_1)$, which is a simple concept. One may remonstrate that this is all very well for a scientist or an engineer who can also deliberately keep $x_2, x_3 \ldots x_n$ constant. Economists could not. It is the old *ceteris paribus/* no-experimentation problem. However, this problem is not necessarily insurmountable in all cases. For instance, if $x_2, x_3 \ldots x_n$ are independent variables, none of which by itself has a marked effect on y_0, and n is large, then, as we vary x_1 deliberately a large number of times, $x_2, x_3 \ldots x_n$ will vary on their own accord in a way which from our point of view is random. It may then be possible with the aid of the central limit theorem to state $y_1 = f(x_1)$ with an acceptable level of significance, or it may be possible to give confidence intervals of y_1 for various values of x_1.[15] I do not want to underrate the problems involved, and in any case either y_1 or x_1 may not be measurable so that statistical methods could not be used. However, a host of everyday business decisions must be based on rough estimates which are in principle of this nature, since business people also are interested in the consequences of their own actions.

The point to be brought out is that we have not chosen to form simple concepts in this way. We have chosen to define our entities such that $y = f(x_1, x_2, x_3 \ldots x_n)$ with a large n, that is, we have chosen to deal with complex phenomena. The complexity is not due to the inherent qualities of the material, but to the definitions we have chosen. The fault does not lie in the material, but with us.[16]

Economists may take comfort from the fact that when the applied physical sciences choose to deal with a complex phenomenon, they are also unable to predict specific events. Applied science cannot even predict the weather very well (and worse, it cannot blame free will for this). The fact that an organizer of an outdoor function cannot be told whether it will rain on a particular night a month hence is in sharp contrast to the fine predictions needed in many industrial processes. However, to blame complexity for this is misleading. Engineers and technicians manipulate the material they deal with, and predict only the consequences of their own actions, not a whole course of events. The meteorologist, on the other hand, is just a helpless bystander taking measurements all over the place; and in this the meteorologist is not unlike the economist turning to the Bulletin of Statistics to do his forecast.

It may be said that we are not always free to choose our own concepts. Just as people take an interest in the weather, so they take an interest in, for example, inflation, and the economist can be expected to say something about it. Even if we had controlled and simple entities at our disposal, we should still have to be able to deal with uncontrolled and complex phenomena. I believe that the approach economists have adopted, as expounded by Hayek, is the correct one, but, as with free will, it does not go far enough. Economists have taken the normal deterministic formula of initial state and governing equations and we have said: 'Let us dispense with the initial state, let us dispense with the actual numerical values of the variables in the equations, and let us simply say something about the form of the equations.' It is an adjustment, but it is the minimum conceivable under the circumstances. We are left with a truncated determinism and this truncated determinism debars us from taking our approach to its logical conclusions.

2.7 The impersonal perspective in the deterministic presupposition

2.7.1 Explanation and prediction

In sections 2.5 and 2.6, I considered two quite different ways in which the failure of the mechanical analogy in economics has been explained. In one (the free-will argument), an appeal is made to our introspective awareness of voluntary action, and the conclusion is reached that human action simply is not determined in a strict way by anything. Whether this introspectively derived conclusion can really be applied to an empirically derived picture of the world is a question that was closely examined by some very eminent thinkers of the past. I shall not go into this question in this work.

The other (the complexity argument) explanation attributes the failure to our limited capacity for grasping the highly complex initial states and governing laws involved, but implies that a deterministic view of human actions is valid in principle. However, though limited, our mental faculties are held to be capable of discovering something, even if very little, about the governing laws – the tenet that leads to what I have called a truncated determinism – and the so-called law of demand is an example of what is meant. The convention of tracing people's actions back, in part, to their tastes or preference functions has allowed some economists, it appears, to reach a compromise between the demands of free will and determinism. In a strictly deterministic view, tastes must have been determined, though they may well have been determined in fields outside the sphere of competence of the economist. It is possible to think of them, however, as changing unaccountably, thus meeting the demands of our introspective awareness, and nevertheless also as data, thus meeting the demands of a deterministic model.

It may seem that it hardly matters for practical purposes whether the strictly deterministic mechanical analogy in economics fails because we cannot obtain the requisite information or because that information does not in fact exist, especially since it seems that a compromise can be found. This view of the matter, however, would overlook the fact that the intention of those who put forward the free-will and complexity arguments is not merely to explain the failure of the mechanical analogy, but to justify their respective approaches to the subject. Since the arguments rest on irreconcilable basic premises, it is not surprising that the methods they advocate are at variance with one another, the basic premise of each invalidating what the other one advocates.

The premise that the human will is not subject to natural law is interpreted to mean that regularities of human conduct are unlikely to be established empirically. We have seen that this leads to a sharp epistemological division between the physical and the social sciences, because a deterministic programme is appropriate in the former, but futile in the latter. However, while the absence of empirical regularities makes it impossible to predict human action, it is of course possible to say something about actions which have already been carried out. They may be interpreted with the aid of the notion of rational action (or of the logic of choice) and of such familiar concepts as desires, intentions, plans, expectations, and so on. The approach to the empirical facts of action is thus to find an intelligible account of past actions, that is, the aim is an explanation of the past, and this is not seen as paving the way for a deterministic prediction of future or later actions. It should be pointed out here that the notion of rational action cannot be derived from empirical data and that it cannot perform the function of a governing law in the deterministic sense (see Chapter 4).

Although the role of rational action in economic equilibrium models sometimes obscures the issue, it should be clear by now that deterministic methods, including the truncated version in economics, require the existence of empirical regularities, that is, regularities which cannot be established by purely conceptual analysis. In so far as use is made of regularities of human conduct, such as the assumption of diminishing marginal rates of substitution or the observation that the income elasticity of demand for most goods is positive, deterministic methods are irreconcilable with the free-will approach, though by no means all economists with a bias towards the free-will argument take the argument to its logical conclusions. Moreover, the proposition that economists should confine their empirical work to explaining past actions does not make sense in a deterministic context, for there the role of empirical regularities is such that the distinction between explanation of the past and prediction of the future is quite unimportant. A closer examination of this issue will be instructive.

I have already mentioned (in section 2.3) that a certain indifference to or diffusion of the observer is characteristic of classical mechanics, the exemplar of determinism. The picture of the universe that Newtonian mechanics has given to the world is such that the observer can be disregarded altogether. If he is present at all, he is omnipresent, and the here and now in which actual observers find themselves is of no particular importance. In other words, mechanics encourages the habit of thought that we can think of ourselves as viewing our subject matter from no particular time or place. It took physicists a long time to realize that this impersonal perspective is not altogether harmless. It seems to me that economists have adopted, particularly in their equilibrium analysis, at least a corollary of this impersonal perspective, namely, the idea that it does not matter from which end we view a sequence of events. If at time t_1 we manage to trace a sequence of events c–b–a back to time t_0, we then feel that we could just as well have started at the other end, at t_0, and have predicted the events a–b–c. Alfred Marshall expressed the idea as follows: 'the explanation of the past and the prediction of the future are not different operations, but the same worked in opposite directions, the one from effect to cause, the other from cause to effect'.[17] This view, it should be noted, follows quite logically from the presupposition that what we see is part of a system (see Laplace's 'système du monde'), and that this system is fully described once its governing laws and its state at any instant of time is known.

I cannot substantiate in a few words my claim that much economic theory is based on the presumption that the operation of explanation is reversible. I hope that subsequent chapters will uncover at least some of the evidence. If my claim may be granted for present purposes, then it is not hard to see that academic discussion between the proponents of free will and of a truncated determinism is unlikely to be very fruitful. As I have pointed out, when members of the Austrian school or others in the free-will camp deal with the empirical facts of human action, they are usually quite content simply to make the past intelligible. The intellectual descendants of the Lausanne school, on the other hand, appear to seek regularities which may serve in deterministic models or, less pretentiously, as guides to action. This, as we have seen, is quite consistent with the deterministic foundations of their neo-classical synthesis, according to which the proper task of the economist is surely to discover something about the equations which govern change in the systems with which the economist is concerned. It may be true that in practice neo-classical economists engage mainly in conceptual analysis, but for this to be considered worthwhile, it must lead to theorems that are testable even if only 'under ideal conditions'. When now the free-will camp contends that the search for regularities is futile, the determinists may well ask what then there is to be done. The answer, that one should explain the past, must surely add to the confusion if at the back of one's mind there is the idea that prediction is explanation in reverse. The basic premises of the respective parties are so incompatible that an exchange of ideas is not really feasible.

2.7.2 Chance and coincidence

I want to suggest that were one to contest the validity of deterministic presuppositions in economics not by invoking free will or complexity, but rather by calling in question the impersonal perspective, one would arrive at a conceptual framework which is more general than either of those underlying the two approaches to the subject I have been discussing. The advantage of this greater generality is that it allows some of the features of both of these approaches to contribute to a common approach. I shall outline such a conceptual framework in section 2.8, while the advantage of its greater generality will be illustrated, I hope, in subsequent chapters. Here, I shall attempt to show that the conceptual framework I have in mind, based on the distinction between *ex post* and *ex ante* facts, would seem quite obvious were it not for the influence of determinism, and more particularly of the impersonal perspective and its corollary that explanation and prediction are the inverse of each other.

In closed models with comprehensive governing equations, all processes are fully determined, that is, calculable, once the state of the system at any instant of time is known. It then does not really matter whether a description of a process refers to something which has already taken place or has yet to take place. However, common sense tells us that in the world at large this is not necessarily so, at least not for us, however it may seem to the vast intelligence Laplace spoke of. Let me take an example. Suppose a stranger to the country asks me to explain the geographical distribution of productive activity in the South African economy. Suppose further that I try to do this by telling a long story which starts with the spice trade between Europe and the East, and features such events as the Great Trek, the discoveries of diamonds and gold, the Second World War, and so on. The enquirer may then be well satisfied with the matter and consider that there is nothing miraculous in the account. Yet the mere fact that the enquirer can look back over a sequence of events and be satisfied in retrospect that the whole thing seems quite credible, does not show that the enquirer could have predicted that sequence of events, or can predict the situation 400 years hence any more than someone 400 years ago could have predicted the present situation. All the enquirer knows is a story built up on countless millions of small events that could have been otherwise – events that were purely fortuitous and coincidental.

No doubt this is quite obvious. It does indicate, however, why explanations of the past usually cannot be inverted. Our explanations of the past invoke chance and coincidence, whereas the prediction of a chance event or a coincidence is a contradiction in terms. If we could predict the economic development of a country in the manner of the Laplacean omniscient intelligence, we would not say that an event happened by chance, or a concurrence of events by coincidence, for we would regard all events as the necessary consequences of 'all the forces by which nature is animated'. The very fact that the notions of chance and coincidence are meaningful to us, sets us apart from that omniscient intelligence, and makes, for us at least, explanation of the past and prediction of the future entirely different matters. The complexity argument is inadequate because it fails to recognize just this. We may well say that the economic development of a country is unpredictable because of its great complexity, and then attempt to devise simple models which approximate some aspects of this development. But this is to recognize the limited capacity of our intellects, without also recognizing that we cannot therefore conceive the world as an omniscient being would, and that we have in fact developed concepts appropriate to our abilities.

We conceive empirical facts in two quite distinct ways. While both conceptions may play a part in explanations, only one is used in predictions. Scientific prediction, whether carried out through the medium of deterministic models or not, is really a statement about, so to say, an *existing structure*, that is, about something which appears to have a continuing (though not necessarily permanent) existence, and may be ascertained in the present. Thus, the prediction of the trajectory of a missile, for instance, is really a statement about the mechanism that propels the missile, about gravity, and so on (see section 2.8). What is important here is that scientific prediction must be distinguished from the statements of fortune-tellers or diviners which purport to be about future events with no evident relation to the present. That explanation may involve a conception of fact quite different from that used in prediction, may be brought out more clearly in those cases in which we can assign probabilities to the occurrence of events. Let us consider a simple case of a lottery in which the winning ticket, to the best of our knowledge, is drawn at random from a total of n tickets. We may then say that the probability of a particular number being drawn is $1/n$. This is a statement about the chance of an event, and not of course a prediction of a chance event. It is prediction in the sense that we are making a statement about the structure of the lottery (random selection from n tickets) in the belief that this structure will persist at least until the draw is made. It is not prediction in the sense that we are saying which number actually will be drawn; such prediction is in the domain of the fortune-teller. If, however, the lottery draw were the subject of an explanation of the past, we would be in a position not only to describe the nature of the lottery but also to relate which number actually was drawn. The chance event may be part of an explanation of the past but not of the prediction of the future. Moreover, our knowledge of the structure of the lottery and our knowledge of which number actually was drawn are not parts of the same operation 'moving in opposite directions', but simply different orders of facts – what I have called *ex ante* and *ex post* facts respectively.

Apart from the given initial states, deterministic models do not admit any facts of the *ex post* type. Being concerned therefore mainly with structures, they allow us to infer the implications of certain changes. It is rather like seeing a part of a machine move and inferring from this what else must be happening in the machine. In this context there is quite obviously a symmetry between explanation and prediction. When stochastic variables are introduced into such models, this symmetry is broken, but if, as is usually the case, these variables are accompanied by probability distributions which are assumed to be known, *ex post* facts still do not really enter the models. As the example of the lottery has indicated, a probability distribution implies a structure of a kind, so that 'pure risk' models usually are also concerned with the implications of structures, like fully deterministic ones. That which has happened by chance or by coincidence is not subject matter of the analysis.

The free-will camp takes very much the same view of 'dead' nature. However, in contrast to the determinist, the basic premise does not admit any structural *ex ante* facts into the analysis of human action. Since men are assumed to act purposefully, the free-will adherent would also deny that human action happens by chance. If a person desires the condition x, his action y may follow quite logically, given that he has the same *ex ante* knowledge z about his physical environment. Since neither the desire for x nor the knowledge z are strictly determined, an observer usually has to infer them from the visible action y. (I shall consider in Chapter 6 the alternative of asking a person about desires and knowledge before the person acts.) To such an observer the action y in a sense does happen by chance, since there is no independent indication of what has led to it, that is, since y is explained by means of the x and z which is inferred from y. If we are uncertain about how an individual will act, we must necessarily invoke chance in explaining the actions in retrospect. With the given state of knowledge (that is, unless an experience reveals new insights into the 'laws of nature') chance and coincidence in the past are corollary of uncertainty in the present. I would maintain therefore that the free-will adherents, unlike the determinists, are concerned with the analysis of *ex post* facts.

My criticism of the free-will argument is that human action is not the only factor which makes the future uncertain for a less-than-omniscient human intellect. The question whether it is in principle the only factor leads straight back to the vast intelligence of Laplace for it is the question whether such an intelligence really can be omniscient, whether the structural analysis of deterministic model can comprehend all phenomena. Those who take a stand on this question will naturally find themselves in either the free-will or complexity camp. However, I do not think that it is worthwhile to pursue such a metaphysical question. We do not need the notion of free will to persuade us that Laplace's vision is far beyond our own experience. To the omniscient intelligence there is no chance, no coincidence and no uncertainty. Explanation of the past and prediction of the future are indeed the same operation. Given the complete system of equations and the state of the world at any instant of time, all things are to him determined to the end of time. Since nothing happens by chance, it does not matter to him whether he views the economic development of a country, the draw of a lottery, or anything else, from before or after it has taken place, from a million years ago or a million years hence. The present here and now in which ordinary mortals must observe the world is of no particular significance to this omniscient intelligence. Nothing could indicate more clearly the metaphysical nature of our deterministic presupposition. It is metaphysical in the very literal sense that this intellect to which we aspire would take us beyond the physical constraints of time and place in which ordinary people, as well as scientists, and even economists forever find themselves.

When we assume perfect foresight in models of perfect competition, we ascribe to the economic subject the abilities of the vast intelligence. We are only too willing to admit that this is an unrealistic assumption, that the economic subject makes decisions in an environment of uncertainty where it is difficult even to assign probabilities to the occurrence of events. But if uncertainty and its corollaries, chance and coincidence, are significant to the economic subject we observe, are they not also significant to us, the observing economists? The mere fact that we aspire to the omniscience of Laplace's vision does not entitle us to analyse our subject matter as though we were omniscient.

2.8 Outline of a sceptic conceptual framework

Let us now leave the notions of determinism and of the dichotomy of dead nature and free will. Let us instead try to think within a different conceptual framework. We shall take the point of view of an observer bound to an eternal here and now. In this situation the observer experiences the continuous flux of events that has intrigued man back to the time of Heraclitus.[18] In an effort to gain empirical knowledge (there is other knowledge), the observer does two things: (1) trying to discern in, or to impose on, the flux of events an order that conforms to the observer's beliefs or presuppositions and thus makes events intelligible to the observer; (2) trying to find in the flux of events some things which seem to endure, or at least to last for some time, and are therefore available to the observer as guides to purposeful action.

I shall enlarge on each in turn. In order to make sense of the flux around them, people draw up an account of the orderly progression of events. This orderly progression conforms to some belief they hold about the nature of things, and in this sense the order is imposed on the events. The account is made up of a number of elements which owe their meaning entirely to their interrelations within the account, that is, the meaning of each element is given by the unique role it plays in the progression of events. These elements are what I have called *ex post* facts. In many contexts it may be necessary to focus attention on the activity of making the past intelligible, rather than on the *ex post* facts which arise from it. I shall refer to this as the *genetic understanding*, not only because it is an adequate description, but because the term genetic is already used in this sense by writers in the philosophy of science.[19] When it is important to emphasize the fact that this understanding is closely related to the human capacity for telling and following a (true) story, I shall refer to it as the *narrative understanding*.[20]

The coherence of the genetic or narrative understanding, as I have said, depends on a set of beliefs. To most educated modern minds an intelligible account of the past must no doubt conform to what appears to be known about causal relations. The further view that everything must have a cause even if the relevant causal relations are unknown, is somewhat different. It makes the task of finding an intelligible account rather easy, since any event that cannot be explained by a known causal relation may then be explained by an imputed one. The advent of a deterministic outlook has probably made the genetic understanding less of a problem than it was at one time. However, it must be emphasized that causal explanations are not essential to this understanding. There may be other arbiters of what is reasonable. Many peoples at many different times, for example, seem to have been quite content to explain what we would call natural events by the behind-the-scenes machinations of all sorts of mythical creatures. Then there are those who, though well aware of the deterministic outlook, would yet maintain that a more profound view of the course of events is attained when one sees Divine intervention in it. Others again may prefer the Marxian dialectical materialism. But if the account that satisfies should have to involve Fate or even gremlins, then this also would do. The important point is that the nature of the belief colours the meaning of the *ex post* facts that arise from the account.

There are not necessarily any lessons to be learnt from the genetic or narrative understanding. This would be so only when an intelligible account cannot be found, so that beliefs and knowledge of structures have to be adjusted to make an intelligible account possible. Such induction, however, is very difficult and does not appear to be the usual reason for trying to make the past intelligible. To the question why therefore people seek a genetic or narrative understanding, I think one must answer that it is vain to look for a reason, as vain as to ask why people seek the approbation of their fellow human beings. It must be accepted as a propensity of the mind, a love for order.

The meaning of *ex ante* facts is easier to grasp because we are familiar with them from the physical sciences. It remains to be shown that when we speak of guides to action, structural facts and causal relations we are essentially speaking of the same thing. I have said that one of the ways in which we try to gain empirical knowledge is to try to discern in the flux of events some things which endure, at least for some time. I want to use a mechanical analogy to illustrate the significance of such an enduring thing. Suppose that I have found a rigid lever which rests somewhere near the middle of its length on a rigid fulcrum, so that end A of the lever rests on the ground and end B is in the air. The lever and the fulcrum, their spatial relationship, my knowledge of the nature of rigid bodies, of gravity, and so on, all this I regard as an indication of a structure, a very particular structure in this case. I also believe that this structure will remain as it is, that is, I regard as unlikely that someone or something will knock the lever over, or that the nature of rigid bodies will change, and so on. The structure is therefore an enduring thing. If now for some reason I wish end A (which is on the ground) to be up in the air, I can press down on end B, and my wish will be fulfilled. I have now used the structure as a *guide to action*. By pressing down one end of the lever, I cause the other end to go up – there is a *causal relation* between the going down of one end and the going up of the other. Moreover, I can *predict* the consequences of my action without actually engaging in it. It may seem trivial when put this way, but I believe that this limited sense of the words cause and predict is the most useful sense, and that it is in this sense that prediction and causation are most usually thought of in the applied physical sciences.

However, mechanical analogies can be pernicious. It is far better to state formal definitions. Since I shall scrutinize certain aspects of economic theory with the aid of the idea of *ex ante* facts in later chapters, great care has to be taken over the definitions, even at the cost of making them rather tortuous. I shall give the formal definitions first and then add some explanatory notes.

> I define a *situation* as events experienced in a particular here and now. A situation contains elements or sets of elements identified by defining procedures. Two or more sets of elements identified by different defining procedures may be related to each other by a relating procedure.
>
> I define *structural knowledge* or a *structural fact* as the belief that if in a particular situation one observes the set of elements A identified by defining procedures B, then by following the relating procedure C, one may observe in the same situation or a later situation specified by the relating procedure C, a set of elements D that can be identified by defining procedures E.

The following are explanatory notes:

1. In the definition of a situation I purposefully avoided using the idea of points in time and space because then, strictly speaking, we could not observe movement. In economics at least we may want to identify, say, a person working, or buying something, and this we must do by judging the intentions which in turn we must infer from the person's actions

which involve movement. The notion of here and now also has other more profound advantages over points in time and space (that is, in this context) which I shall not go into here.

2. By defining procedure, I mean the way in which we subsume aspects of situations under classes, that is, how we identify something as, say, water, a dog, the barking of a dog, houses, a person offering something for sale, exchange transactions, and so on.

3. It is important that the elements observed and the elements predicted be identified by quite different defining procedures. Such is the case with, say, the aural impact of a dog's bark and the visual impression of a dog. The statement that there must be at least one buyer and one seller whenever a sale takes place, is not a structural fact. It is a purely analytical statement based on the meaning of the words used. Only one set of defining procedures is used and the statement merely explores its implications. Many statements in economics are of this type.

4. Following David Hume, who was of course a pioneer of this kind of conceptual framework, one would expect relations to be either spatial or temporal. Relating procedures (which were not part of Hume's set-up) would then be of a type such as look around the corner, or wait ten minutes, and so on. In economics, however, one may require far more intricate relating procedures. Note that relating procedures relate classes of elements and not unique elements in individual situations.

5. Structural facts may be expressed as mathematical functions or relations, provided two additional requirements are met. First, it must be possible not only to subsume elements of situations under classes by defining procedures but also to relate these elements uniquely to all possible members of their respective classes by measurement procedures. Second, it must be possible to state the relating procedure mathematically. The definition of a structural fact is therefore more comprehensive and may be used in contexts where there are only generic concepts.

6. The relation between a set of elements observed and the set of elements predicted may be asymmetrical. If x is a sufficient condition for y, y is a necessary condition for x and the relation is not reversible unless y is also a sufficient condition for x. The definition of a structural fact is for the case where the observed elements are a sufficient condition for the predicted ones. (For example, if I stand next to the wall of a house and hear the bark of a dog without seeing a dog, I shall believe that if I look around the corner, I shall see a dog.)

One could rephrase the definition to serve other cases. For instance, let x stand for a sufficient condition for y, and let y stand for a necessary but not sufficient condition for x. Then the observed absence of y is a sufficient condition for the absence of x. The observed presence of y, on the other hand, creates a certain potential for x (where potential is used so that probability may be reserved for something that can be given a numerical value) and y may here be called a potential condition for x. For instance, if I see that there is no dog in the hotel room next to mine, I shall believe it unlikely that I shall be disturbed by barking from that quarter (that is, absence of necessary condition). Whereas, if I see that there is a dog there, I shall believe that there is a certain potential for being so disturbed (that is, presence of potential condition). The relation between sets of elements observed and predicted may also be symmetrical, where each is a sufficient and necessary condition for the other, or even where each is merely a potential condition for the other.

7. It must be emphasized that the definition of a structural fact is not a definition of a structure. It would be difficult to define structures. They must be inferred from structural facts and many such facts may be stated about what we feel intuitively is a single structure. By following up various structural facts we may be able to pinpoint where a structure must lie.

We must now make the transition from structural knowledge to the closely related knowledge of causal relations. Let us consider two cases of purposeful action. (a) The captain of a ship waits for high tide before leaving a harbour. (b) An industrial chemist produces chlorine from rock salt or brine in order to manufacture a household bleach. In both cases a knowledge of apparent structures is used as a guide to action. The two cases differ in so far as in (a) there is no question of the captain having *made* the tide rise *so that* he could leave the harbour – the tide would have risen in any case – while in (b) the chemist had to *make* the salt yield chlorine *so that* he could manufacture the bleach. The structural fact of the succession of tides is of the type I defined earlier. It can be seen from this definition that we may come by such knowledge as purely passive observers. We assert a structural fact when we simply notice some regularity like the succession of tides or like the constant conjunction (to use Hume's terminology) of barking noises and dogs. The sort of knowledge applied by the chemist in (b), on the other hand, is of a type that is unlikely ever to spring to the mind of a passive observer. Salt does not break up into its constituent elements of its own accord; we must make it happen in order to see that it is possible.

It may be instructive to visualize whatever structures there may be in such a way that it appears that some are revealed to passive observers by events around them while others are forever hidden to passive observers. It seems to me that by far the greater part of the knowledge revealed to us by the physical sciences pertains to such hidden structures. One may ask how one can find structures that are hidden. The answer, it seems, is that the technique and the principle we follow is that we must somehow activate hidden structures in order to see that they are there at all. When we observe a certain set of elements in a situation, we perform a deliberate operation which brings about a new situation which contains a new and predictable set of elements. The new element that we introduce into the course of events we may define to a great degree of accuracy since we create it, and it is between this element and the aspects of the emerging situation towards which our interest is directed that we say there is a causal relation.

> I define *knowledge of causal relations* or *causal facts* as the belief that if one performs the operation X in a situation in which one observes the set of elements A identified by defining procedures B, then by following the relating procedure C, one may observe in a later situation specified by the relating procedure C, a set of elements D that can be identified by defining procedures E.

The similarity between this definition and the one for a structural fact is obvious, as in that case, one could elaborate the definition by distinguishing between sufficient and potential conditions. This would introduce the idea of a potential cause and in some cases of a measurable probability. I indicated this in the shorter version of the definition at the end of section 2.3, I also indicated there that one could allow for a number of distinguishable stages between the operation X and the appearance of the set of elements D. This would introduce the idea of a closed system, or of a dynamic model of a process. It is the deterministic dream that such models may be expanded to include all events – a dream which does not seem to allow for anyone outside of the closed system to activate the whole thing. These deterministic ideas, it seems, also lead us to speak of lines of causation that somehow enter the model through stochastic events or even to suppose that causation is vaguely represented by the notion of equilibrating forces. What this overlooks is that the outcome D (in my definition) depends as much on the set of elements A as on the operation X. The line of causation is established purely by what we take an interest in. It is a convention of thought to say that there is a causal relation between the element introduced by the operation X and the outcome D. Of course, I am not saying that structures are established by our deliberate operations. Whatever structures there are, presumably would be there whether we acted or not. 'But to think of a relation between events as causal is to think of it under the aspect of (possible) action.'[21]

Since the structural and causal facts that I have defined are so similar, I shall also use the more comprehensive term *ex ante facts* to cover them both.

With the definitions out of the way, let us make sure that it is well understood that *ex ante* and *ex post* facts are simply quite different. Because an *ex ante* fact seems to retain its character in a variety of situations, we may rightly regard it as a distinct entity with some kind of separate existence. An *ex post* fact, on the other hand, is not an entity distinct from the context in which it appears and cannot be said to have had a separate existence at any time. It is merely part of an intelligible account of the course of events.[22]

Let us take stock now in order to see what sort of picture emerges from this conceptual framework. At the same time let us bear in mind the free-will and complexity arguments that I discussed in sections 2.5 and 2.6. Confronted by incessantly changing events, we try to bring order to these events by finding an account of their progression that is intelligible to us because it conforms to our beliefs. We also seek knowledge that indicates the existence of structures that are permanent or at least last for some time. We may seek such knowledge of structures merely to satisfy our curiosity, or we may do so to use it as a guide to action. But when we use it as a guide to action, we use it for the very limited purpose of predicting some of the consequences of our own actions, that is, we want to make our purposeful actions effective. We may seek to frame our structural and causal knowledge in ever more general and comprehensive terms, but we do not build it up into a vast model that could predict the entire course of events. We may occasionally mistake our intelligible accounts of the past for such vast models, but all the same, as far as the human intellect is concerned, the entire course of events remains ever unpredictable. And if this is all that some of us want to show (and there may be more to it), then we do not need the notion of free will; all we have to do is to recognize the humble nature of the human intellect.

2.9 The selection and testing of hypotheses: two aspects of the problem of induction

The sources for our empirical work can only be the present situation and our records of the past. It may therefore be argued that a genetic understanding of events that can only be achieved if we make adjustments to our beliefs is exactly that mental activity of induction by which we ascertain *ex ante* facts, and that our direct experience therefore always relates to *ex post* facts. It is a profound truth that we cannot have a direct experience of *ex ante* facts, and it is this which is usually referred to as the problem of induction.

There is a vast literature on this problem and any study that sets out to suggest means by which *ex ante* facts may be established in economics would have to tackle this literature. In this work, however, I am concerned only with what I regard as a confusion between *ex post* and *ex ante* facts in economics, and I have therefore steered clear of the problem of induction and shall continue to do so. For reasons that I shall give in what follows, there are, however, a few remarks about the problem of induction that I should like to make here.

One may distinguish two aspects of the problem of induction. First, there is the question of validity of *ex ante* facts, or the justification we may have for believing in a hypothesis. The first clear statement of this aspect of the problem was given by David Hume.[23] Second, there is the question of a practicable method of inductive inference where there is not much prior knowledge, that is, the question of the selection of hypotheses to be tested when there are not many leads to go by. I shall refer to these two aspects as the question of validity and of approach respectively.

I mention this because I think it is relevant to an objection that may be raised against my suggestion of a possible confusion between *ex post* and *ex ante* facts in economics, and especially against the further suggestion that as a result of this confusion no really serious attempts appear to have been made to isolate true *ex ante* facts. Am I not aware, it may be asked, that empirical work in economics is dominated by regression analysis and the testing of hypotheses, that is, by clear attempts to isolate structural and causal knowledge? Now, I have no doubt that regressions and significance tests are a very valuable aid for finding constant conjunctions and causal relations when these are not obvious to the eye, provided that there is some prior reason for believing that they may be there. The proviso is important, as I shall try to explain.

Opinions may differ on the value of the results achieved by regression analysis in economics. It would be futile to argue about what is of any consequence and what is not. But if anything is amiss, then one could have a shrewd idea that the fault does not lie in the mathematical part of the statistical procedures but rather in the selection and collection of the data before they reach the mathematical stage. In other words, it seems more likely that the shortcomings of our inductive methods, if any, hinge on the question of approach rather than on the question of validity.

These remarks apply equally to the wider issue of falsifying hypotheses, however the tests are conducted. The falsifiability criterion, in the form given to it by Karl Popper, has attained what one could call a measure of popular appeal among economists. But economists appear to have adopted it with the idea that it is a complete method of induction, one that provides an answer to both aspects of the problem of induction. Popper himself, however, expressly disclaims any interest whatever in the question of approach and concerns himself purely with the question of validity. I shall quote rather fully from the work in which he develops the falsifiability idea, so that his attitude may be seen clearly.[24]

> The initial stage, the act of conceiving or inventing a theory, seems to me neither to call for logical analysis nor to be susceptible of it. The question how it happens that a new idea occurs to a man – whether it is a musical theme, a dramatic conflict, or a scientific theory – may be of great interest to empirical psychology, but it is irrelevant to the logical analysis of scientific knowledge. The latter is concerned

not with questions of fact, but only with questions of justification or validity. Its questions are of the following kind. Can a statement be justified? And if so, how? Is it testable? ... In order that a statement may be logically examined in this way, it must already have been presented to us. Someone must have formulated it and submitted it to logical examination.

Accordingly, I shall distinguish sharply between the process of conceiving a new idea, and the methods and results of examining it logically.

Some might object that it would be more to the purpose to regard it as the business of epistemology to produce what has been called a 'rational reconstruction' of the steps that have led the scientist to a discovery – to the finding of some new truth. But the question is: what, precisely, do we want to reconstruct? If it is the processes involved in the stimulation and release of an inspiration which are to be reconstructed, then I should refuse to take it as the task of the logic of knowledge. Such processes are the concern of empirical psychology but hardly of logic. It is another matter if we want to reconstruct rationally the subsequent tests whereby the inspiration may be discovered to be a discovery or become known to be knowledge.

However, my view of the matter, for what it is worth, is that there is no such thing as a logical method of having new ideas, or a logical reconstruction of this process. My view may be expressed by saying that every discovery contains 'an irrational element', or a 'creative intuition', in Bergson's sense.

Popper is of course entitled to delimit his field of interest as he pleases. But it must not be forgotten that 'falsifying hypotheses' is at best only half an inductive method, and the second half at that. If one were to take it seriously as a whole method, then even so simple a problem as diagnosing why the air in my office seems rather stale would be a gigantic task. If my 'creative intuition' was not very sound, I would have to take hypotheses at random from that infinite number that it would be possible to devise. I would have to try relating the staleness of the air to everything from the phases of the moon to the state of the stock market, and I would not finish the job in a lifetime. There must be some way of narrowing down the field of likely hypotheses, and in practice, of course, there is.

That many economists do not really take the full implications of falsifiability to heart was brought out by the controversy that followed the publication of Professor Friedman's paper on 'The Methodology of Positive Economics'.[25] Friedman maintained, quite correctly in the context of falsifiability, that the validity of hypotheses must be judged by their implications, that is, their predictive capacity, and that 'to suppose that hypotheses have not only "implications" but also "assumptions" … is fundamentally … wrong and productive of much mischief". In the ensuing controversy there was much discussion of the 'realism' of assumptions and of the need that theories should pass this test too. But this steps out of the ambit of the principles behind the falsifiability criterion completely. After all, what are the assumptions of a hypothesis? Surely, only the hypothesis itself. And how do we judge the realism of the assumptions? The test is whether the implications remain unfalsified. The matter really is quite simple if one takes falsifiability to heart. However, many economists found it irksome (not to mention a certain suspicion that Friedman had an axe to grind) to have to accept hypotheses, however absurd they may otherwise seem, simply because they have not been falsified when put to the test. They wanted to apply to hypotheses the kind of once-over by which we judge whether the fellow across the dinner table is talking sense or not. And this is not merely a minor addition to the falsifying procedure. It establishes a completely different principle. The additional criteria for selecting hypotheses presumably cannot be established by taking hypotheses out of the blue and putting them to the falsifying test, for these surely would also have to pass the realism test, and one would be back at the same problem, multiplied a few times over. What Friedman's critics seemed to be getting at is that we need not only a method that reasons from the unknown general to the known particular, as Popper has described the testing of hypotheses, but also a method that starts from what we already know.

In everyday problem-solving we do not pick hypotheses out at random. We use a host of consciously known or subconsciously recognized structural and causal facts to narrow down the field of possible hypotheses, before ever we put a hypothesis to the test. When I open my office window to remove the stale air around me, I may not be successful. The stale air may be outside, or whatever. But I have good reasons for preferring the hypothesis that relates the stale air to the position of my window over one that relates it to the phases of the moon or the state of the stock market. Again, one may diagnose an engine failure within minutes if one knows how car engines work; if one does not, one probably would not care to posit any hypotheses at all. The precedence that the question of validity has taken over the question of approach in much methodological discussion is probably due to the fact that the physical sciences were seen as the exemplar, and here so much is already known that the question of approach takes care of itself. But the physical sciences presumably also had to be built up from fundamental facts that no one would question. Who could fail to notice, for instance, the regular alternation of day and night or of winter and summer?[26] But what of fields where constant conjunctions are not so common. When one is confronted by a bewildering flow of events that no 'creative intuition' seems able to explain, then surely the question of approach should take precedence over that of validity. After all, falsifying hypotheses is all very well when one hardly ever succeeds, but not much fun when one succeeds every time. Whatever Popper may think, we do need a logical reconstruction of the way knowledge is built upon knowledge and a criterion for what sort of knowledge we can begin with.

In economics we naturally do have certain conceptions of the sort of knowledge we should begin with, and we frame our theories accordingly. When empirical work calls for hypotheses, we do not pick them at random but select them according to these conceptions and theories, and then judge their validity by well-known methods. However, when one considers questions such as the inflation of our time or the transfusion of a more production-oriented culture into subsistence economies, then perhaps one may wonder whether we really know where to begin. When everyone has had their say, do these questions not remain as enigmatic as ever? At least those who think they do may consider it worthwhile to examine again the approach aspect of the problem of induction, as distinct from the aspect of logical justification, that is, the selection as distinct from the testing of hypotheses. It is towards such a possible examination that I hope the distinction between *ex post* and *ex ante* facts may make some small contribution.

2.10 The philosophy of science applied to economics

I realize that exercises in the philosophy of science applied to economics are not to everyone's taste. Also, it may be held against me that I could quite easily have outlined the conceptual framework I am putting forward, and have had my say on hypotheses, without ever mentioning the word determinism. However, not only was it necessary to give some indication of why a different conceptual framework may be justified, but it will be necessary in the following chapters to refer back to the nature of determinism. When I consider certain aspects of the history of micro-economic theory in the light of the concepts developed in this chapter, it hardly would be a promising tack to assume that by some strange coincidence the eminent thinkers of the past were not able to muster what may seem like common sense to the person in the street. It will be necessary to view their work in terms of what most likely were their philosophical presuppositions.

Of course, it would not be necessary to consider presuppositions at all if our science were daily revealing new truths of great intellectual beauty or of great beneficence to mankind. But when the existing paradigms have been explored down to almost the last niche and still sensible people feel there is much room for improvement, then we cannot afford to ignore the philosophy of science. I should like to quote one more passage from Sir Arthur Eddington's lectures, because it indicates the potential role of a philosophy of science applied to economics. In discussing how the intrusion of philosophy into the new physics of the 20th century created discomfort among the physicists of his day, he says:

> This vagueness and inconsistency of the attitude of most physicists is largely due to a tendency to treat the mathematical development of a theory as the only part which deserves serious attention. But in physics everything depends on the insight with which the ideas are handled before they reach the mathematical stage.[27]

Notes

[1] Pierre Simon, Marquis de Laplace, *A Philosophical Essay on Probabilities*, translated from the 6th French edition by F.W. Truscott and F.L. Emory (New York: Dover, 1951) p 4. This work originated in a lecture course delivered in 1795, published as the 'Essay' in 1814 and incorporated, as the introduction, into his great *Théorie analytique des probabilitiés* of 1820. The original French version may be inspected in *Oeuvres Complétes de Laplace*, tome septième (Paris: Gauthier-Villars, 1886) p vi.

[2] *Ibid.*, p 6.

[3] Physics made significant new progress when this conception of the universe began to be questioned in the 20th century. For instance, we are told that when Einstein reasoned that simultaneity at a distance is unobservable, he was led to his special theory of relativity of 1905. See, for example, Sir Arthur Eddington, *The Philosophy of Physical Science* (Cambridge: Cambridge University Press, 1939) especially chapter III, or Michael Polanyi, *Personal Knowledge* (London: Routledge, 1958) pp 9–15.

[4] See, for example, P.W. Bridgman, *The Logic of Modern Physics* (New York: Macmillan, 1927) especially pp 209ff. Bridgman wanted to promote the idea of the operational meaning of concepts, so perhaps he had an axe to grind. There can be no doubt, however, that he was a leading physicist.

[5] Eddington, *op. cit.* (note 3) p 63. The book was based on lectures he delivered as Tarner lecturer of Trinity College Cambridge in 1938.

[6] *Ibid.*, p 90, his italics.

[7] I shall return to this point in later chapters. I hope to show there that a careful distinction between *ex post* and *ex ante* facts could free equilibrium analysis from deterministic presuppositions.

[8] In an earlier version of this text, this paragraph featured as the first paragraph of section 2.7 of this chapter.

[9] Many of Shackle's numerous works treat the question. A full exposition is to be found in G.L.S. Shackle, *Epistemics and Economics* (Cambridge: Cambridge University Press, 1972).

[10] G.L.S. Shackle, 'Decision: The Human Predicament' *The Annals of the American Academy of Political and Social Science*, March, 1974, p 2. I am indebted to Professor Lachmann for pointing out this publication to me.

[11] I shall deal with some of the views of Menger and von Mises in Chapters 4 and 5, and I shall give detailed reference there. Von Mises's most comprehensive exposition of economics as a branch of praxeology is to be found in *Human Action* (London: Hodge and Yale University Press, 1949) while his views on the proper study of the proper study of action as empirical fact may be gathered from *Theory and History* (London: Jonathan Cape, 1958). Further expositions and development of the aspects of the Austrian approach under discussion here may be found in L.M. Lachmann, *Capital and its Structure* (London: Bell, 1956) especially the chapter entitled 'On Expectations', pp 20–34; also in his *The Legacy of Max Weber* (London: Heinemann, 1970) especially pp 5–7 and 29ff: also in his paper 'Die

geistesgeschichtliche Bedeutung der Österreichischen Schule in der Volkswirtschaftslehre', *Zeitschrift für Nationalökonomie*, XXVI, 1966, pp 150–67.

See also I.M. Kirzner, *Competition and Entrepreneurship* (Chicago: University of Chicago Press, 1973), especially chapters 1 and 2, where he argues that the concept of entrepreneurship cannot be properly understood within the confines of a strictly deterministic model. F.A. Hayek's association with this school has become a fairly loose one.

[12] Von Mises's views need special mention here. On page 25 of *Human Action* he says: 'There are for man only two principles available for a mental grasp of reality, namely, those of teleology and causality.' On page 22 he had said: 'The archetype of causality research was, where and how must I interfere in order to divert the course of events from the way it would go in the absence of my interference in a direction which better suits my wishes?' (I take this to be another version of the definition of an ex ante guide to action that I gave at the end of section 2.3.) Subsequently, von Mises speaks of 'the causal methods of the natural sciences' (p 27). Let us suppose that there was a person who knew von Mises's viewpoint well and also his intentions to promote his brand of method. If this person had wanted to provoke von Mises to disagreement, he could have lauded behaviorism; if he had wanted von Mises to think he was a sensible fellow he could have ridiculed behaviorism. In both cases he would probably have stood a good chance of success. Would von Mises have agreed that this person would have been using causal knowledge outside of the physical sciences? On the basis of what von Mises said in a later work, which I want to consider in Chapter 4, one could probably come to the tentative conclusion that he would have agreed.

[13] An earlier version of this chapter had the following text in the place of the last paragraphs of this section:

> Now, it is obvious that free will and a determinate nature must be juxtaposed for the notion of free will to make any sense at all, just as the juxtaposition of free will and divine guidance was necessary in the original theological context. However, the question remains why the dichotomy has to be made at all. After all, most of what the proponents of the free-will argument say would make equally good sense if determinism were abandoned altogether. Their doctrine would be affected only in so far as the rationale for their rejection of the methods of the physical sciences in the social sciences would disappear. The claims that methodological individualism can make the past intelligible to us by referring to the meaning people attached to their actions, would remain unaffected. But there would now be no a priori reason for believing that the physical-sciences type of search for guides to action could not also bear fruit in the social sciences provided that it is accepted that determinism is not a prerequisite for such a search in the physical sciences either.
>
> I do not know whether the proponents of the free-will argument stopped short of denying determinism altogether because the case for determinism just seemed too plausible or because the fundamental distinction between the social and physical sciences had to be defended at all costs. (Of course, there may have been no reason at all.) Having seen that the deterministic presupposition introduces difficulties into economics, they reduce its scope, but they are not prepared to abandon it altogether.

[14] F.A. Hayek, 'The Theory of Complex Phenomena' in *Studies in Philosophy, Politics and Economics* (London: Routledge, 1967) pp 22–42. The article first appeared in 1964. It is preceded in the book by an article which deals with the same subject. Entitled 'Degrees of Explanation', it first appeared in 1955.

[15] If the variables are not independent probability theory usually cannot help us. In this connection, Hayek mentions a paper by Warren Weaver (in the article cited in note 14). Weaver distinguishes between organized and disorganized complexity and asserts that the

biological and social sciences have to contend with the former, where probability theory can be of little help. In 'A Quarter Century in the Natural Sciences', *The Rockefeller Foundation Annual Report*, 1958, Weaver seems to take it for granted that economic entities, such as markets, must be seen in the same light as biological organisms. It is, of course, possible to formulate economic models in such a way that many of the variables are dependent, and there may well be some similarities between economic equilibrium and homeostasis (that is, for example, the fact that mammals maintain a fairly constant temperature). But the *prima facie* case for treating a supposedly equilibrating economy as an organism does not seem to me very strong. It would be different if we found, say, that the price of bread is always at a certain level or in a certain relation to all other prices. But this is not so. I hope to deal with this issue on another occasion.

16 It is an epistemological commonplace that subject matter and concept are not independent of each other. We do not choose concepts to fit the facts. The choice of concepts and the cognition of the fact is the same thing. On p 2 of the book cited in note 14, Hayek features Goethe's remark to the effect that everything factual is already theory. (Karl Mittermaier wrote in hand writing at the top of the page where this footnote is found: 'Das Höchste wäre zu erkennen, dass alles Faktische schon Theorie ist' ['The highest is to recognize that everything factual is already theory']).

17 Alfred Marshal, *Principles of Economics* (8th edn, London: Macmillan, 1920) p 773. Marshall was not the first to express the idea and in fact he was less than unequivocal about it. The quotation appears in a discussion of the role of induction and deduction in economics. He agrees with J.S. Mill that economic forces combine in the manner of the forces of mechanics, but then turns to a version of the complexity argument. '[T]he forces of which economics has to take account are more numerous, less definite, less well known, and more diverse in character than those in mechanics; while the material on which they act is more uncertain and less homogeneous.' He then goes a step further: 'Again the cases in which economic forces combine with more of the apparent arbitrariness of chemistry than of the simple regularity of pure mechanics, are neither rare nor unimportant.' Noting that even chemists deal with a less intractable material than do economists, he comes to a quite undeterministic conclusion: 'The function then of analysis and deduction in economics is not to forge few long chains of reasoning, but to forge rightly many short chains and single connecting links.' The idea of isolated *ex ante* facts having thus been accepted, the sentence I have quoted comes as a surprise and one may feel its deterministic implication is unintentional. However, a few lines further on he returns to the argument, this time adding Laplace's proviso of omniscience. 'While in so far as our knowledge and analysis are complete, we are able by merely inverting our mental process to deduce and predict the future almost as certainly as we would have explained the past on a similar basis of knowledge.' Marshall's adoption of partial equilibrium analysis is consistent with his wavering on this issue.

18 Heraclitus of Ephesus lived in the 6th and 5th centuries BC. Sir Karl Popper, *The Open Society and its Enemies* (London: Routledge, 1945) vol 1, pp 9–14, credits Heraclitus with an immense influence. It was 'the genius of Heraclitus', he says, 'who discovered the idea of change'. 'The view he introduced was … that the world was not a more or less stable structure, but rather one colossal process: that it was not the sum-total of all *things*, but rather the totality of all events, or changes, or *facts*' (his italics). Popper continues: 'The philosophies of Parmenides, Democritus, Plato, and Aristotle, can all be appropriately described as attempts to solve the problems of that changing world which Heraclitus had discovered. The greatness of this discovery can hardly be over-rated. It has been described as a terrifying one.' Popper's interpretation of Heraclitus is more extreme than many others. Heraclitus did have an explanatory model. Needless to say, it tried to explain the flux of events.

[19] For example, Ernest Nagel, *The Structure of Science* (London: Routledge, 1961) p 25: 'The task of genetic explanations is to set out the sequence of major events through which some earlier system has been transformed into a later one.' Nagel says that 'it is moot question whether it constitutes a distinctive type', and implies that the doubt arises from the fact that causal explanations are used in these genetic explanations. I shall argue in what follows that this is not necessarily so. The term 'causal-genetic' is also in use, but is quite unsuitable for my purposes.

[20] See W.B. Gallie, *Philosophy and the Historical Understanding* (London: Chatto, 1964). Gallie examines at length what it means to follow a story. I have avoided the word 'historical' because history would bring with it a whole new set of presuppositions. Also, some historians seem to me to prevaricate on the distinction I am trying to make, sometimes implying that there are lessons to be learnt from history and, when pressed on this point, taking refuge in the idea that history is after all unique.

[21] G.H. von Wright, *Explanation and Understanding* (London: Routledge, 1971) p 74. Von Wright deals at great length with the issue under discussion here. I shall quote a few more passages from this book to lend support to my contentions. 'I would maintain that we cannot understand causation ... without resorting to ideas about doing things and intentionally interfering with the course of nature' (p 65). 'It is established that there is a causal connection between p and q when we have satisfied ourselves that, by manipulating the one factor, we can achieve or bring it about that the other is, or is not, there' (p 72). 'The idea that man, through his action, can bring about things is founded on the idea that sequences of events form closed systems. ... The identification and isolation of systems again rests on the idea that man can do things, as distinct from bringing them about, through a direct interference with the course of events' (p 68). 'The illusion has been nourished by our tendency to think ... that man in a state of pure passivity, merely by observing regular sequences, can register causal connections and chains of causally connected events which he then by extrapolation thinks pervade the universe from an infinitely remote past; to an infinitely remote future. This outlook fails to notice that causal relations are relative to fragments of the world's history which have the character of what we have here called closed systems' (p 82). It must be remembered that we are here concerned with a conception of causes and not with the way the word 'cause' is actually used. Languages are notorious for making a single word serve many purposes.

[22] The reader is invited at this point to reread Popper's description of the terrifying discovery of Heraclitus, which I quoted in note 18. One could interpret Popper as saying that Heraclitus discovered that there are only *ex post* facts and that these are not things, as structures would be, but merely facts.

[23] David Hume, *A Treatise of Human Nature*, Book I, 1739 (many editions are available). It was probably Hume's greatest achievement to have pointed out this logical problem (though modern linguistic analysis claims to have solved it). He had an immense influence on later generations, as well as on his contemporaries, not least among whom was Kant. The influence of Kant's attempts to solve the problem has manifested itself in economics, through various channels, in the *a priori* logic of choice of von Mises and Lord Robbins. Through different channels it has influenced the free-will argument. A notable book, in the tradition of inductive logic following Hume and the earlier work of Bacon, was J.S. Mill's *A System of Logic* of 1843. Book VI of this work, entitled 'On the Logic of the Moral Sciences', is especially relevant to economics. (Moral sciences was Mill's term for what we now call social sciences.)

[24] K.R. Popper, *The Logic of Scientific Discovery* (London: Hutchinson, 1959; original German edition 1935). The extracts are taken from pp 31 and 32.

[25] Milton Friedman, 'The Methodology of Positive Economics' in *Essays in Positive Economics* (Chicago: University of Chicago Press, 1953) pp 3–43. See also the discussion

of falsifiability in economics in M. Blaug, *Economic Theory in Retrospect* (2nd edn, London: Heinemann, 1968) pp 666–75. Blaug also gives an extensive bibliography of the controversy over the realism of assumptions in economics.

26 The capacity for seeing constant conjunctions and for jumping to conclusions, however, may differ considerably. For instance, *The Star*, Johannesburg, 6 December 1973, reported the case of a man in England who, clad in a mauve suit, raped two women on successive Thursdays and in both cases stopped for a drink in a bar afterwards. The police thereupon requested all publicans in the area to look out for a man in a mauve suit on the following Thursday. On the other hand, there has been a report of a small New Guinea community whose members apparently are not at all aware of any connection between coitus and childbirth.

27 Eddington, *op. cit.* (note 3) p 55.

Structure and Equilibrium

3.1 The line of investigation

In the remaining chapters I shall try to use the conceptual framework outlined in Chapter 2 to investigate certain aspects of the method of comparative statics in micro-economic theory. I hope that this will serve the dual purpose of casting some new light on micro-economic theory and of illustrating and enlarging upon what I have said in Chapter 2. This chapter will prepare the ground for the further analysis.

The applicability of the *ex post ex ante*, genetic structural, distinction is in no way restricted to micro-economic comparative statics. I think it could be very usefully applied to macro-economics but this would require first a rather difficult conceptual analysis of what I called defining procedures in the definition of structural and causal facts. Growth and other dynamic models could also prove to be a fertile field, since deterministic ideas appear to play a more prominent part in these than they do in comparative statics. I have chosen micro-economic comparative statics because I want to investigate some rather basic concepts, and for these purposes micro-economic comparative statics is the least complicated part of economic theory. While the distinction between statics and comparative statics will be seen to be an important one, it will obviously not be possible to deal with the comparison of static equilibria without also dealing with statics.[1]

3.2 Causal facts in practical applications

Though the term comparative statics is not as old,[2] the notion goes back at least to Pareto, Marshall and even Cournot. In broader terms, it was also the method in, for instance, Ricardo's 'Essay on the Influence of a Low Price of Corn on the Profits of Stock' of 1815. Since there was opposition to the Corn Laws in the England of his day, it was after all natural that Ricardo should ask himself what would be the consequences of their abolition, or, if he advocated their abolition, that he should want to show others what these consequences would be. In this sense the method of comparative statics is simply a search for guides to action, or for causal facts as I have defined them in Chapter 2. (The reader is invited to look at the definition again to convince himself of this.)

The ordinary Marshallian demand curve, if it could be established as a fact, would be a medium for expressing causal facts, for it could answer the vital question: What would happen if the price or the quantity marketed were changed? No doubt Marshall intended his analysis of demand and supply to be available as a guide to action even in the form in which he offered it to his readers, as is shown for instance, by his illustration that a tax levied on a commodity that 'obeys the law of diminishing return' raises the price by less than the amount of the tax.[3] One does not have to look far in the *Principles* to find similar examples. Book V specifically deals with effects of changes which may be analysed in terms of demand and supply. However, Marshall does not make much of the distinction between changes that we bring about deliberately and changes that we merely observe to be taking place, that is, of the distinction I have made between causal and structural facts.

The observation, that an increase in the demand for X is accompanied by a rise in the price of X, is different from the tax case; for simplicity I shall leave the elasticity of supply out of account. Of course, if it is a case of increasing the incomes of consumers and of knowing that they will spend some of this increase on X, then we are still concerned with a causal fact, though it would have been better to express it by saying that the operation by which we increase consumer income raises the price of X. The passive observation, that whenever we see an increase in demand under certain conditions we may also see an increase in price, may seem like a constant conjunction or what I called a structural fact. But it is not. How do we observe an increase in demand? If we argue that a particular rise in price shows us that the demand must have risen (possibly because of a change in tastes), and that the rise in price is due to this rise in demand, then this is certainly not a structural fact as I have defined it. For that definition (section 2.8) I stressed that the sets of elements (A and D) that one relates (by C), must be identified by quite different defining procedures (B and E). This is not the case here; we have identified only a rise in price. However, I am here anticipating some of the difficulties I want to consider. There is much to be done before they can be considered more rigorously.

3.3 Deterministic notions and axiomatic constructs in equilibrium theory

While the attempt to isolate causal facts is often quite evident when micro-economics is applied to the analysis of some practical problem, it is not so evident in the more formal statements of micro-economic theory. Here one finds a mixture of deterministic notions and purely axiomatic constructs. Let us see how Professor Samuelson has explained the rationale of his analysis. The following quotations and page numbers come from his *Foundations of Economic Analysis* (Harvard University Press, 1947). On page 8 he said:

> This method of comparative statics is but one special application of the more general practice of scientific deduction in which the behavior of a system (possibly through time) is defined in terms of a given set of functional equations and initial conditions. Thus, a good deal of theoretical physics consists of the assumption of second order differential equations sufficient in number to determine the evolution through time of all variables subject to given initial conditions of position and velocity.

He then pointed out that this applies equally to partial and general equilibrium systems: the scope of the latter is simply wider. 'The things which are taken as data for that system happen to be matters which economists have traditionally chosen not to consider as within their province' (p 8). Among these are 'tastes, technology, the governmental and institutional framework, and many others'. Since 'there is nothing fundamental about the traditional boundaries of economic science' and 'a system may be as broad or narrow as we please depending on the purpose at hand' (p 9), one may presume that even tastes or the institutional framework may be regarded as variables whose solution values we wish to find. So far the picture is entirely deterministic. However, the paucity of information restricts the economist to a truncated determinism.

> In the absence of complete quantitative information concerning our equilibrium equations, it is hoped to be able to formulate qualitative restrictions on slopes, curvature, etc., of our equilibrium equations so as to be able to derive definite qualitative restrictions upon the responses of our system to change in certain parameters. (p 20)

On page 7 he had already said that 'our theory is meaningless in the operational sense unless it does imply some restrictions upon empirically observable quantities, by which it could conceivably be refuted'. (The word 'qualitative' in this context refers to changes in quantities of known direction but unknown extent.)

The question then arises how one may arrive at 'meaningful theorems' about the qualitative restrictions on solution values as parameters are changed. One way is to postulate such restrictions in the equilibrium equations themselves, as when diminishing returns are postulated. Samuelson, however, concentrates on two other sources of theorems. One is the assumption that the behaviour of consumers and firms may be regarded as maximizing (or minimizing) behaviour, so that theorems may be derived from the mathematical extremum conditions. The other is the assumption that an equilibrium is stable. ('How many times has the reader seen an egg standing upon its end?' he asks in support of such an assumption.)

In regard to the latter of these, Samuelson made a particular contribution. Let us first remember that comparative statics is a 'special application' of the deterministic scheme he has described. It is special, apparently, because one investigates 'changes in a system from one position of equilibrium to another without regard to the transitional process involved in the adjustment' (p 8). While the complete transitional process may be disregarded, Samuelson showed that the assumption of the stability of equilibrium even 'in the small' as he calls it (p 262, that is, in the neighbourhood of equilibrium or for small displacements), presupposes a theory of dynamics, 'namely a theory which determines the behaviour through time of all variables from arbitrary initial conditions' (p 260). He illustrates this by pointing out the differences between the so-called Walrasian stability conditions, those in Marshall's period analysis and the asymptotic approach to equilibrium in the cobweb analysis (pp 263–8). Thus, Walrasian stability implies that price always rises with an increased demand but quantity may rise or fall. Marshallian long-run stability implies that quantity always rises with an increased demand but price may either rise or fall. This difference in the qualitative restriction is due to a difference in outlook based on implicit dynamic models. Samuelson took the study of the stability of equilibrium further than, for instance, Hicks had done in *Value and Capital*, and he announced his well-known 'correspondence principle', that is, 'that there exists an intimate formal dependence between comparative statics and dynamics' (p 284). In other words, a study of dynamic models may yield information with which to compare static equilibria.

If my interpretation of him is correct, Samuelson feels that the choice between deriving theorems from the assumption of maximizing behaviour (maximum and minimum problems) or from the assumption of stability based on some postulated dynamic model, is a matter of practical convenience. It depends on what kind of assumptions one can reasonably make and on how these may best be brought into equilibrium models. Moreover, the two sources of theorems are not unconnected. Maximizing considerations are at the back of, for instance, the stability of an equilibrium market price. This, again, fits in with his view that static equilibria are 'simply degenerate special cases' (p 285) of dynamic systems.[4]

All in all, therefore, Samuelson managed quite well to put his work into a deterministic mould. He even went as far as to say (p 9) that the 'existence of such systems' does not depend on symbolic or mathematical methods and that any part of economic theory 'which cannot be cast into the mold of such a system must be regarded with suspicion as suffering from haziness'. And yet he does not quite seem to have been able to keep it up himself, for immediately after saying this he made a few remarks that do not seem to be at one with the deterministic interpretation. He proceeded to point out that the relationship between variables in equilibrium systems is one of mutual interdependence, and that 'once the conditions of equilibrium are imposed, all variables are simultaneously determined'. In the next few pages he put this a little differently. The derivation of the equations stating the solution values of unknowns from those giving the functional relationship between variables and parameters may involve difficult mathematical calculations, but the former are nevertheless logical implications of the latter. We merely 'bring to explicit attention certain formulations of our original assumptions' (p 12) which may be tested. These thoughts prompted him, it seems, to remark that (p 9) 'it is sterile and misleading to speak of one variable as causing or determining another', and that it is only 'as a figure of speech' that a change in a parameter can be said to cause a change in a variable. He thus arrived at a view very much like the undeterministic view of causation reached by modern analytic philosophy which I tried to explain in section 2.8 of Chapter 2. Once cause and effect have been removed, not much can remain of a deterministic scheme, and it was then that he expressed very neatly the quite undeterministic way in which many mathematical economists in fact seem to regard equilibrium. He said: 'Indeed, from the standpoint of comparative statics equilibrium is not something which is attained: it is something which, if attained, displays certain properties.'

I have pointed out this ambiguity in interpretation because I think it illustrates my contention that one finds a mixture of deterministic notions and axiomatic constructs in the formal development of micro-economic theory. I want to suggest that this ambiguity has something to do with the fact that one tries to give an 'economic meaning' to the mathematical properties of equilibria. Let us look at this a little more closely. When one asks for the 'economic meaning' of some theorem, or of some other deduction from a model, one usually means that one would like it to be put in a form in which one can visualize or picture it. (The exceptions are cases when it can be converted into a theorem in the logic of choice – see Chapter 4.) The words 'visualize or picture' are important in this context. A proposition acquires 'economic meaning' when we can conceive it in terms of the concepts with which we are familiar from our visual or other sensory experience. Thus Walras's analysis of an equilibrium market price conjures up a vision of buyers and sellers haggling in a market, or of agents shouting out prices, if the market is well-organized in Walras's sense. Marshall's long-run picture focuses attention on the entrepreneur maximizing profits under certain market and technical conditions which we can also visualize. The deterministic interpretation of equilibrium, the mechanical analogy and even the word equilibrium also are attempts to allow us to visualize vaguely how one thing leads to another and how various forces operate and may reach a balance.

While it may often be possible to formulate such schemes mathematically or to put such an interpretation on a mathematical computation, mathematical methods certainly do not need any visualizing to be meaningful on their own terms. I think it is generally agreed that mathematics is an *a priori* science in the sense that it does not need experience or observation, that is, that it may be developed without any sensory contact with things external to the mind. The *a priori* axiomatic method of mathematics therefore not only does not require us to visualize the proof of a theorem, but in many cases, as even the founder of general equilibrium theory realized, such visualizing is simply not possible.[5] We may well visualize, though not very realistically, a consumer weighing up marginal rates of substitution and equating their ratios, to price ratios, but the 'meaningfulness' of maximum calculations does not depend on such pictures. If very much the same idea is stated in the form that a consumer reaches a point in a multidimensional commodity space, we cannot visualize what he is doing. Mathematics, as a hypothetico-deductive science as it has been called, is concerned with axioms which define relations between non-specific terms, and with the logical demonstration that certain theorems are implied by the axioms. Whether anything in experience conforms to the axioms and therefore to the theorems is another matter. The difficulty of making a deterministic picture fit an axiomatic construct hinges on a distinction which goes back to Aristotle, namely, between the temporal order of experience and the logical order among proposition.[6]

It appears that in the work of, for example, Debreu, Arrow and Hahn on competitive equilibria and allied fields, the notion of equilibrium as an axiomatic construct has come to the fore. At the same time deterministic pronouncements have become rare, and dynamics and comparative statics, on which Samuelson placed such great emphasis, have receded into the background. On comparative statics, Arrow and Hahn came to the conclusion that 'the kind of parameter changes for which predictions become possible is pretty limited'. On Samuelson's idea that meaningful theorems may be derived from two sources, they say: 'In fact, very few useful propositions are derivable from this principle.' Their book does not devote even one chapter specifically to dynamic systems.[7]

Equilibrium, then, is something which displays mathematical properties and these necessarily must be expressed in *a priori* terms. If equilibria were to display properties in our sensory experience, we would have to think of it as something which lasts at least long enough for us to investigate its properties, and it would have to have a certain resilience to the prodding of the investigator, like an organism. Alternatively, we could think of it as an explanation of the past which we have frozen at a particular stage so that we may analyse certain relations. Neither of these interpretations fits the mathematical conception of equilibrium. It is not an empirical thing or fact, but an axiomatic construct. I am not suggesting that general equilibrium theorists would be inclined to justify their work by saying that they have found mathematics to be fun. Certainly, the axioms or assumptions are phrased in terms of ideal types drawn from everyday economic experience, and the verbal comments contain images every bit as sensual as Marshall's fish markets, as when the displacements which can be withstood by local and global stability respectively are likened to the burning down of either one or 'half the factory'.[8] But the question of how and where in experience we are to find something that conforms to general-equilibrium axioms is left rather vague, which is not to say that these axioms may not one day be discovered to have great practical application.

Professor Hahn seems to have had such difficulties on his mind when he delivered his inaugural lecture at Cambridge.[9] Commenting on the fact that the philistines do not appreciate the *a priori* purity of general equilibrium, he says: 'And indeed it is a fair question whether it can ever be useful to have an equilibrium notion which does not describe the termination of actual processes' (p 8). He points out that the Arrow-Debreu equilibrium 'makes no formal or explicit causal claims at all' (p 7) and that the weak causal claim, that any actual economic process can terminate only in an Arrow-Debreu equilibrium, would be false (pp 10–11). After explaining why he believes that the notion of static equilibrium nevertheless has its uses, he turns to the tendencies in his own thought. Our equilibrium notion, he feels, 'should reflect the sequential character of actual economies', and in such a way that it cannot be reformulated non-sequentially – it must be 'sequential in an *essential* way' (p 16, his italics). With the temporal order of experience an essential part of it, the notion must be conceivable in experiential terms, and one can certainly visualize the notion of equilibrium which Hahn then proceeds to outline tentatively.

It hinges upon the idea of learning but learning in a special sense. In my terminology, it is the acquisition of *ex ante* facts, but not that of *ex post* facts. Hahn merely illustrates the difference (p 19). An agent assigns probabilities to the two events that it will and will not rain in Cambridge at t+1. When he has experienced t+1, he will of course know whether it actually rained. This increase in knowledge is not learning. He would learn only if the event t+1 affected his weather forecasts from then onwards, that is, in my terminology, if he acquired structural facts. An agent's *ex ante* facts make up his 'theory' (p 18 – Hahn puts this differently, of course). Agents also receive 'messages from the economy and nature'. Since an agent has certain motives and a theory, the messages prompt certain acts. The relation between messages and acts is the agent's 'policy' (p 20). On page 25 he comes to his tentative definition of equilibrium: 'an economy is in equilibrium when it generates messages which do not cause agents to change the theories which they hold or the policies which they pursue'.

When one considers this vision of equilibrium one may see that it has a deterministic feature. One must take into account that it is a very tentative vision, that Hahn had certain ideas about types of agents which he did not want to discuss and that he stresses 'the difference between the perceived environment and the environment' as such (pp 24–5). Nevertheless, this vision of equilibrium seems to require us to visualize messages which actually *cause* people to change their knowledge and perhaps even their motives, for equilibrium is a state in which no such messages are generated – and this looks suspiciously like the sort of picture to which Shackle objected in the passage I quoted in section 2.5 of Chapter 2. It is surely not Hahn's intention to say that we are always in a state of equilibrium because people's thoughts are as original as Shackle believes them to be. So this illustrates again that attempts to devise constructs that one can visualize, and that are therefore of relevance to empirical work, have tended to lead to deterministic notions. In this case they are introduced because Hahn realizes that the notion of equilibrium has to be supplemented by empirical constants or regularities in order for it to be useful.

> What is more important in my present context is that it is precisely the empirical claim for the usefulness of the equilibrium notion that the theories and motives of agents are sufficiently stable and that we are not allowed to involve changing theories or motives to help us out of falsified predictions. (p 23)

He also indicates where one should look for such stability. He says that 'certain institutional environments only permit certain kinds of behaviour to qualify for equilibrium behaviour' (p 23) and that 'equilibrium actions of agents will reveal themselves in habitual behaviour' (p 23). Now, this is in many respects the point of view which I want to promote by the present study. But I do not think we can develop it with deterministic notions. A distinction such as that between *ex post* and *ex ante* facts is needed.

One could extend the analysis of this section to the works of other economists, but this would no doubt become extremely tiresome. The points that I have tried to illustrate in this and the previous section are these: Causal and structural facts, as I have defined them, enter intuitively into many practical applications of micro-economic theory, and may even play a small part in the work of a theorist. However, when the formal and the axiomatic constructs of micro-economic theory are given economic meaning, deterministic notions have tended to appear. They may be proclaimed boldly or they may creep in surreptitiously, but always sufficiently, it seems, to prevent analysis within a consciously sceptic conceptual framework such as I have suggested.

3.4 Behavioural equations and *ex post* and *ex ante* facts

I shall now try to show the link between conventional equilibrium models in economics and the sceptic conceptual framework which I am proposing. In order to do this, I shall put forward a three-fold system of classifying data. The aim will be to show that certain features of economic models correspond to this classification of data while *ex post* and *ex ante* facts also correspond to it in certain respects. The system of classification is thus an intermediary and may later be dropped in favour of the *ex post* and *ex ante* distinction. However, since the demonstration that, say, an assumed institutional arrangement conforms to my definition of a structural or causal fact may in some cases be an extremely laborious task, I shall also use the three-fold classification of data in section 3.5, as well as on several occasions in later chapters.

The question is perhaps best approached in the following way: A mathematical equilibrium model usually incorporates behavioural equations which set out the institutional, technological and other assumptions made. Let us say that the function $y=f(x)$ is such an equation. This is a very general statement which says no more than that there is some relation between the independent variable x and the dependent variable y. We may regard $y=f(x)$ as an assumption which we may use together with other assumptions to derive theorems that follow logically from them. We would then be concerned with what I called axiomatic constructs in the previous section, that is, with the logical order among assumptions or axioms. On the other hand, we may believe that $y=f(x)$ is something we could observe, or the existence of which we could infer from observation via the Samuelson-type meaningful theorem and the falsifiability criterion. In that case, $y=f(x)$ would have an empirical meaning and we could say that it expresses an inkling of the existence of a structure linking x and y. I shall assume that we are dealing with functions that have empirical meaning.

In $y=f(x)$, both the independent variable x and the relation indicated by the symbol f are non–specific data. Both may take on various specific values or forms, that is, both may change. Since both are data, the independence of x cannot distinguish it from f. In an empirical context one may distinguish between them on the basis of duration. The independent variable x must be thought of as changing more rapidly than f. The f is a rule for mapping or transforming the set x into the set y, and however rapidly f is changing (that is, taking on different specific forms and values), x and y must be changing more rapidly, if f is to retain its meaning as a rule. If both f and x were changing equally rapidly, there would be no function between x

and y. If now for function or rule we read structure, we may come to the conclusion (again in an empirical context) that the cognition of a structure requires the prior cognition of something that is changing.

Moreover, if we could specify $y=f(x)=ax$, a further distinction on the basis of duration would be introduced. The rule for transforming the set x into the set y would now be partly specified. It is 'multiply by a' whereas it could well have been 'add a', or have involved terms other than a, have been quadratic, and so on. A mathematical operation has been specified and in this context it must be regarded as a specific structural datum. It is also, so to say, a constant, for in the form that the function now has it does not change, whereas the parameter a may still take on various values. However, the parameter a must also be distinguished on the basis of duration from the independent variable x, for otherwise we would have written $y=f(a,x)$ and not $y=f(x)$. In an empirical context one may therefore distinguish between constant, parameter and independent variable on the basis of the duration of their specific values.

As we shorten the period under consideration, a parameter must eventually take on a specific value, such as when we write $y=2x$. But this is the shortest period for which the notion of structure is meaningful, and it is as far as the specification of a structure can go. When we shorten the period even further and solve for a specific value of x, then we are not specifying a structure but a particular situation, or two particular situations, in which, say, $x=3$ and $y=6$. The independent variable x is a datum but not a structural datum. As already noted, we have to be aware of a changing x before the relation f can have a meaning, and this is simply another way of saying what I stressed in relation to *ex ante* facts (section 2.8, Chapter 2), namely, that structures retain their character in a variety of situations.

In an empirical context, therefore, a constant is a specific structural datum, a parameter a non-specific structural datum, and an independent variable a non-specific datum. They are distinguished by the duration of specific values which is unlimited in the case of a constant and shortest in the case of an independent variable.

The classification of data that I have in mind is based on this tripartite distinction. However, before this system of classification may be defined, there is one other matter that needs consideration. We do not always conceive the institutional framework of an economy in a way in which it can, or can readily, be put in the form of numbers or mathematical operations. It is, therefore, desirable that we think of the data that we want to classify as sets whose elements may either differ quantitatively or qualitatively (where the word qualitative has its more usual meaning and not the meaning it has in Samuelson's 'qualitative restrictions'). If they differ quantitatively, the sets are variables and the elements their specific values. If they differ qualitatively, the sets are classes and the elements their specific forms. In philosophical language, it is desirable that the system of classification should be able to handle not only universal-particular concepts but also generic concepts.

With this in mind, we may now define the three classes of data as follows:

1. Casual data are data sets whose elements, when found in experience, are treated as though they will be found for only a moment, where moment is defined as a period too short for purposeful action to be executed. (In other words, the specific values or forms of casual data are changing continuously, or too fast and uncertainly to be useful as guides to action.)
2. Parametric data are data sets whose elements, when found in experience, are treated as though they will be found for periods longer than a moment, where moment is defined as before (that is, they may be used as guides to action).
3. Institutional data are specific elements of data sets found in experience which are treated as though they will last indefinitely.

It may be seen that what are differences of specification in behavioural equations have here been translated into differences of duration. In $y=f(x)=ax$ the transformation rule has been partly specified, that is, multiplication has been specified while the parameter a has not. In terms of this classification, one would say that multiplication is institutional and 'a' is parametric. Of course, the degree of specificity is not much of a problem when one is concerned with rational numbers and mathematical operations, for a term is either symbolic or specific. In most empirical work, however, complete specification, if not inconceivable, is at least not feasible and the degree of specificity may be very important. One could have drawn up a system of classification that distinguishes data on the basis of both duration and specificity, but this would have required a detailed description of what I called relating and defining procedures in the definitions of *ex ante* facts, and such a description would be in itself a larger work than the present study.

The demonstration that *ex post* facts correspond in some respects to casual data and *ex ante* facts to parametric and institutional data, will be much more brief because the correspondence is in fact quite close. The notion of structure developed here is that of a functional relation (though it is not essential that the relation be a function), and we have seen that this notion of an empirical structure implies a duration longer than that of the specific values or forms of casual data. An *ex ante* fact was defined as the belief that certain things may be observed whenever certain conditions are met or whenever certain operations are performed when such conditions are met (section 2.8 of Chapter 2). It is therefore a belief in an empirical relation, and as such it is excluded from the class of casual data. The function $y=f(x)$ expresses an *ex ante* fact, and x in y loosely correspond to the sets A and D respectively in my definitions of structural and casual facts. (The sets x and A are casual data). Since no distinctions on the basis of duration were made between different *ex ante* facts, they may fall into either the parametric or institutional classes of data.

It would be far more significant to carry the link between conventional behavioural equations and *ex ante* facts on to our practical apprehension of economic institutions. This cannot be done without the additional conceptual analysis I mentioned earlier, but a vague indication of this further link may here be in order. I would ask the reader to make a distinction between analytical definitions of, say, banking systems, markets or private property in the abstract, on the one hand, and the empirical description of a particular banking system, a particular market or the particular norms relating to private property in a particular situation, on the other. The distinction is not clear-cut. It is a matter of degree. I believe it can be shown that the empirical description of a particular institution must consist of an indefinite number of what are in principle causal or structural facts. (By empirical description I do not mean an historical account of how an institution has evolved. This would of course contain *ex post* facts.) Two very prosaic examples may be mentioned. The operation of making out and delivering a cheque under certain conditions has certain consequences (causal fact). Banking hours are constantly conjoined with certain clock and calendar times (structural fact). The reader may like to demonstrate to herself – or try to falsify – that her descriptions of particular markets or of particular norms relating to private property, are always in terms of *ex ante* facts (remembering though, that it is always possible to draw up analytical definitions without much recourse to observation). As I have said, such demonstrations are sometimes very laborious, and I shall not attempt them in what follows. I shall, however, classify an economic institution as institutional data, or as parametric data when it is thought of as changing sporadically, and not as a single datum.

It remains to be shown that casual data in some respects resemble *ex post* facts. We have seen that the cognition of an empirical structure requires the prior cognition of casual data. If we wanted to find $y=f(x)$ by induction, we would require the prior knowledge of both x and y. Given the function or structure, one of the two may be derived from the other, but one (and by convention x) cannot be explained or predicted by the structure. In the absence of other information, casual data therefore come about or happen by chance, at least as far as we are concerned. However, there may be other information. A variable that is exogenous to one system or structure may be endogenous to another, and it may therefore be possible to build up ever larger structures. If $y=f(x)$ and $x=g(z)$, then $y=h(z)$. Thus, the long chains of causation of deterministic systems are built up by linking ever more *ex ante* facts. But always there are two loose ends. The one end is of course the situation we wish to explain or predict. The other end we must eventually call the initial state, that is, we must either accept it as having come about by chance, or, what amounts to much the same, we must find some explanation for it that does not depend on a structure, as when we say that the initial state happened to be the wish of an omnipotent deity, or even the desire of a very mortal man. The explanation of the genesis of a situation I have called the genetic understanding and the specific values or forms of casual data may now be seen to be *ex post* facts abstracted from their context in the genetic understanding.

3.5 Some enquiries into the data of economic models

It has now been shown, I hope, that there is a link between the behavioural equations of conventional economic models and the *ex post* and *ex ante* analysis. One may rightly ask why I have subjected the reader to such an arduous demonstration. Can a sceptic conceptual framework bring anything to light that is not equally accessible by conventional methods? I hope to persuade the reader in the remaining part of this study that it may well do so. At this stage I merely want to broach an issue which I shall here probe from various angles, and one aspect of which I shall examine more closely later on. The issue arises out of the subject matter of the previous section. Broadly speaking it is this: What general considerations may be brought to bear on the classification of the (usually assumed) data of economic models as casual, parametric or institutional? Are the conclusions one may reach in this way reflected in the behavioural equations used? Each of the following exercises in sceptic analysis will briefly probe one or other aspect of these questions. The last will draw various threads together.

3.5.1 Static equilibrium without casual data

Let us take up the argument which I used towards the end of section 3.4. By linking *ex ante* facts, a deterministic model may trace a chain of causation from an initial state to some situation in which we are interested. There are then two loose ends, the initial state and the final state (z and y in $y = h(z)$). Now, it appears that one may avoid initial states and the final states altogether if one can somehow tie up the two loose ends in a structural relation. In that way a closed system of relations would be formed in which all variables are interlinked, or mutually determine each other. Something like this seems to have been done in equilibrium theory. Let us consider the equilibrium of a very simple two-commodity pure-exchange economy. The model consists of six equations, namely, four behavioural equations relating the quantities demanded and supplied of each good to the prices of both goods, and two equilibrium conditions. In general form, the equations are:

$$D_1 = f_1(P_1, P_2)$$

$$S_1 = g_1(P_1, P_2)$$

$$D_2 = f_2(P_2, P_1)$$

$$S_2 = g_2(P_2, P_1)$$

$$D_1 = S_1 = Q_1$$

$$D_2 = S_2 = Q_2$$

Since in equilibrium

$$f_1(P_1, P_2) = g_1(P_1, P_2) \text{ and}$$

$$f_2(P_2, P_1) = g_2(P_2, P_1)$$

one may solve for the prices P_1 and P_2, and therefore for the quantities Q_1 and Q_2, if the parameters are known and the functional forms are such as to make solutions possible. The exogenous variables P in the behavioural equations have been made endogenous by the equilibrium conditions and the assumption that one is dealing with a state of equilibrium. There appear to be no casual data and no initial state; the whole thing is one structure of interlinked relations.

What is the empirical status of such a structure? Let us first consider two extremes between which deterministic models with initial and final states may be said to fall. At one extreme one regards all phenomena as casual data. It is the 'terrifying' view Heraclitus had of a cacophonous flux of events (see notes 19 and 20). At the other extreme there is only structure. The vision of one grand structure standing in monumental silence may seem even more terrifying, but it would be incorrect to describe it as a vision, for it is not something that we can visualize. Casual data, we have seen, are a prerequisite for a structure that we can visualize, that is, one with an empirical meaning. When casual data are eliminated we can no longer visualize a structure. In deterministic models the initial state brings about the final state – something still happens – and we can visualize the intervening structure. But when we create systems of mutual interrelationships, we eliminate not only the initial state but also the final state. Nothing could ever happen in a world of pure structure. If anything happened in the everyday sense of this word, there would necessarily be casual data.

None of this need detract from the equilibrium concept if we do not try to visualize it, and accept it as a system of logical interrelations, an axiomatic construct which, in Samuelson's phrase, displays certain (logical) properties. Above all, this concept of equilibrium is timeless so that the assumptions made for it cannot really be classified under any of the categories I have put forward.

3.5.2 Static equilibrium in a temporal context

The modern concept of economic equilibrium was arrived at by more or less deterministic means. Walras, for instance, developed his equilibrium of a pure exchange economy from a picture of traders with certain initial endowments and certain ideas about what they wanted to have. Equilibrium then had to be hit upon by some *tâtonnement* mechanism, which Walras did not consider fanciful in 'well-organised' markets with brokers.[10] It seems that many economists have not allowed equilibrium to slip from such beginnings into an atemporal, non-visualizable form. Static equilibrium is not timeless, not even a state of complete rest, but rather a version of the stationary state in which things keep on churning over in the same old way. It is a state in which there are no incentives for anyone to do anything except to repeat what was done before, not because there is nothing further one wishes for, but because, under the constraints of one's resources, of technology and the institutional environment, this is what one most prefers to do.[11] There is therefore movement but no change. Though unlikely, this equilibrium is visualizable.

Let us imagine stationary-state statisticians who record the sales turnover of each commodity in every equal time interval t, and the price of each commodity at the end of every t. The respective figures for each commodity in every t are always the same, and so of course is the ratio between the corresponding figures for any two commodities. The statisticians are displaying some of the logical properties of the state of equilibrium in which their community lives. However, their figures cannot give them the slightest indication of the behavioural equations which we, the economists looking in from the outside, know to be the determinants of their equilibrium. In order to reveal any of the behavioural equations (such as the relation between price and the quantity supplied) the statisticians' prices and flows would at the very least have to change – but they are always the same.

Let us now suppose that there is a change in a parameter of one of the behavioural equations, and that the system adjusts itself instantaneously to the new equilibrium. The statisticians will be startled to find that their figures and the ratios between them suddenly are all different. Something actually has happened in the sense in which things just happen in our everyday non-stationary-state experience. Let us suppose further that the frequency of parameter changes increases so that there is at least one change in every t (or in every sub-interval of t to which the statisticians, breaking their stationary state condition, may turn for their recordings). The statisticians, like their real-life counterpart, finds that nearly all their figures change from period to period and they are not at all aware of any equilibrium. Yet all these changes do not necessarily enable them to find the relevant behavioural equations, for they do not know in each case whether a change is due to a move along a curve or to a move of a curve itself. As far as they is concerned all their data are casual data and they are as bewildering as the flux of Heraclitus. However, we, the economists looking in from the outside, could tell them that they are in fact experiencing a whole series of static equilibria.

The question that presents itself is whether we really understand the matter any better than the bewildered statisticians. The difference between their and our view seems to be that they see mere changes whereas we see changes in parameters, that is, changes in structures. But how do we know that they are changes in structures? Do we not need other data that are changing much faster if we are to come to such conclusions?

I shall leave for later consideration the question of whether from a sceptic point of view the concept of static equilibrium can be said to provide a useful insight into economic phenomena. I should however, mention that I have not been trying to suggest here that the answer is necessarily 'no'.

3.5.3 Comparative statics and ceteris paribus assumptions

In comparative statics we assess the effects of a change in one of the data while all the other data are held constant. The *ceteris paribus* assumption involved here may be a purely heuristic device or it may reflect an empirical assessment of the structure of an economy. When a question is investigated for the first time, the *ceteris paribus* thought experiment is no doubt a useful device for dealing with one thing at a time. However, if *ceteris paribus* assumptions are used in comparative statics as an expedient for avoiding assertions about what are the structural facts of an economy, comparative statics may become a rather barren and uninteresting bit of logic.

Let us take a case where the various influences on the quantity demanded are considered equally variable, but where nevertheless one speaks of a demand curve as a relation between a set of prices and a set of quantities. (As we have seen, this would not really be very sensible since the parameters of the demand function would have no meaning; all the same it is done in practice.) Say that at a particular time the demand curve could be expressed as $y=30-2x$ (where y is the quantity demanded and x the price) and that $x=3$. If x now changes to 5 and at the same time the parameters also change, let us say to $y=100-8x$, then it would be quite correct but rather pointless to say that, though the quantity demanded actually increased by 36 units, *ceteris paribus* – if the parameters had not changed – it would have decreased by 4 units. Again, and in more graphic terms, anyone who steps off a busy city pavement into a building may say that, *ceteris paribus*, there is now one less person on the pavement, or that there is one less than there would otherwise have been. It would be impeccable logic, but it would not tell us whether there would then be fewer or more people on the pavement. I do not think that economists normally intend the method of comparative statics to be an exercise in such empty logic. When we say that, *ceteris paribus*, the imposition of a tax will raise the price of x, we mean that usually it will do so, that the circumstances under which it will not do so are somewhat exceptional. To be useful as a guide to action, the method of comparative statics requires us to commit ourselves to an empirical judgement about which are the institutional or parametric and which the casual data in a particular problem.

Unfortunately, such empirical assessments are usually not written into economic models, and sometimes the latter may be quite misleading in this respect. For instance, on looking at demand and supply functions separately one may come to the conclusion that price and quantity are the casual data and the relations between them the parametric data; after all, that is the way the functions are drawn up. However, when the two types of function are combined in the analysis of market equilibrium, price and quantity are no longer data at all. If, as often happens, we take a deterministic view and regard the equilibrium price and quantity as the final state in which we are interested, then we should sort through whatever data remains to see which is the initial state (because there must be one if there is a final state) and which the governing laws. In other words, we have to make up our minds about which are the casual and which the parametric or institutional data. The casual data, it appears, have been included, in the model in the guise of parametric data.

3.5.4 The demand curve

Let us consider a Marshallian market demand curve (standard or Friedman's interpretation) as a datum or data of a model. Where then does it fit into the threefold classification of data that I developed in section 3.4? Clearly, it has never been regarded in such a way that one could classify it as an institutional datum. But is it a casual or a parametric datum, that is, is a particular curve valid for only a moment, in the sense in which I have used the word in this context, or for longer than a moment?

One may rephrase this question in more familiar terms. The demand curve holds *ceteris paribus*, and the question then is whether the 'other things' are casual or parametric data (or a mixture of the two). Professor Friedman considers the question in his well-known article on the Marshallian demand curve. He says that the other things are not the 'same over time' but the 'same for all points on the demand curve'.[12] The points on a demand curve are 'alternative possibilities' not 'temporally ordered combinations', which

I take to mean that the curve is a functional relation between a set x and a set y. A function in itself, of course, is atemporal, but since the other things are not to be seen as the same over time, the time curve presumably must be seen as valid for only a point in time. When the demand curve is taught at an elementary level, the question of its durability is often left rather vague. In his popular elementary textbook, for instance, Samuelson merely says of the demand curve: 'Thus there exists, *at any one time* a definite relation between …'. The phrase that I have put in italics seems to suggest frequent variation, but of course it actually says that demand is not intermittent, and nothing about how long a specific demand function may be expected to last. Even the 'law of gravity' may be said to exist at any one time. Later in the book Samuelson says: 'The demand curve is drawn on the assumption that these other things do not change' and then poses the question: 'But what if they do?'[13] Marshall himself was far more explicit on this point. However, I shall leave that over for consideration in Chapter 5.

Let us have a closer look at the *ceteris paribus* assumptions. Friedman presents a list of five 'other things' in Marshall's work and argues that it is a fairly comprehensive list.[14] It is (with inverted commas for Friedman's text):

1. 'purchasing power of money'
2. 'amount of money at his command'
3. 'custom'
4. price of 'a rival commodity'
5. range of rival commodities available.

The standard interpretation of the demand curve, he claims, ignore items 1 and 5 and extends 4 to all commodities. The only reasonable interpretation of customs, he says, is that it refers to tastes and preferences (about which more later) and item 2 he takes to refer to income and wealth.

Purely heuristic devices are not in keeping with the spirit of Marshall's analysis. How can he then account for these *ceteris paribus* assumptions without prejudicing the practical usefulness of partial equilibrium in comparative statics? My own opinion is that he appealed to common sense and urged his readers to develop an acumen for whatever regularities there may be. However, there is something more definite to be said. There is, for instance, what Hicks has called a simplification of genius, namely, the assumption that a consumer spends only a small part of his resources on any one commodity. This implies that a change in the price of the commodity under consideration, and isolated changes in other prices, have a negligible effect on the consumer's 'willingness to part with money', that is, the marginal utility of money may be considered a constant, or, in Hicks's terminology, the income effects of price changes are negligible.[15] In the present context, one may say that Marshall made the institutional assumption that there are many commodities and therefore many prices and partial markets. (Since Marshall did not like the idea of rigidly defined commodities one should really say that he assumed that one could take this view of the economy.) The constancy of the marginal utility of money must then hold for the demands of the majority of commodities, but not necessarily for all commodities, and not if prices move in concert.

Friedman disagrees with such an interpretation because Marshall speaks of a demand curve for, for example, wheat and houseroom. He argues that Marshall must have assumed that price changes are accompanied by compensating variations in the prices of unrelated commodities so that the marginal utility of money remains constant. As far as I can see, the only explanation of a structural nature that Friedman advances for such a relationship between prices is Marshall's quantity theory of money, which also needs the heuristic assumption that the quantity of money is constant.[16] Friedman also speaks of an 'organising principle' and of analytical devices whereby the purchasing power is kept constant. From a purely structural point of view, therefore, there may be more to be said for the standard interpretation, but this does not of course have any bearing on the correctness of either of them since this would require a judgement on the relative importance Marshall attached to inductive generalizations and the consistency of theories.

Even so, Marshall's simplification of genius would account for only item 1, the purchasing power of money, on the list of *ceteris paribus* assumptions, and for the income effects generated by price changes in the extended version of item 4. But changes in the prices of substitutes (and of complements for that matter), and in the available range of substitutes, can be expected to affect the quantity of a commodity demanded, quite apart from any income effects. It is hard to imagine that one could ever justify the inclusion of these in the '*ceteris paribus* pound' on any grounds other than as a heuristic device. The prices of commodities cannot always change more rapidly than those of substitutes and complements (which are involved in the parameters of demand functions) even if only because the relationship would have to be reversible. One may see, therefore, even without considering income and preferences, that the parameters of the demand functions in partial equilibrium analysis contain data which in a temporal context cannot be considered parametric.

3.5.5 Demand in an exchange economy

Ceteris paribus assumptions are less troublesome in general than in partial equilibrium analysis simply because there are fewer 'other things' that have to be kept constant. The problematic prices of substitutes and complements in partial analysis, for instance, simply vanish from the list of data. It may be interesting therefore to take up the question of the data of demand again in the simple case of the general equilibrium of a pure exchange economy.

It may be taken as understood that the neo-classical type of analysis always traces demand back to the individual and that it is there seen to depend on two entities, namely, income and preferences. In the pure exchange economy, with which explanations of general equilibrium analysis very often start, income takes the form of initial endowments. An initial endowment consists of the commodities an individual happens to possess at the time that is chosen as the starting point of the analysis. During the course of the analysis, individuals exchange commodities in order to reach an equilibrium allocation. Initial endowments therefore exist for only a moment in the initial state and, in this simple conception of an economy, they are casual data while preferences are parametric data because, though changeable, they are the same in the initial and final states. There are also other implicit data, such as property norms and market institutions, which are treated as though they are a permanent feature of the economy and are therefore institutional data.

This is how Walras saw the matter in the earlier parts of his analysis, and it is also the system which is represented by the manna-gatherer economies found in some textbooks.[17] It should be noted that the parameters of the individual demand functions are here partly dependent on casual data and on the initial state. *Ex post* facts somehow appear in the guise of *ex ante* facts.

3.5.6 Some irreducible data

Since we considered the contrast between the stationary-state statistician's view from the inside and the economist's from outside, in section 3.5.2, some evidence has come to light to suggest that the view of the statistician may have been the more correct, since economists include casual data in their parameters. However, both partial equilibrium and the general equilibrium of an exchange economy are somewhat artificial cases in which some of the endogenous variables of a broader general equilibrium system have to serve as data. One has to turn to such broader systems in order to assess the data that are irreducible from the economic point of view. I shall attempt to do this here. However, the Walrasian general equilibrium system is altogether too complex for an easy assessment of the nature of the data involved. I shall therefore try to simplify it. The procedure for simplification will be the following: I shall consider the institutional data in a particular formulation of an economy and then seek a more general formulation in which the institutional data may be reclassified as parametric data. In other words, I shall try to find more general conceptions of economies in which the more particular conceptions are special cases which may last for some time.

Let us formulate a very general data set whose elements are various forms of affiliation among individuals. We may therefore speak of a set of affiliations or, in order to fit in with certain developments in economic theory, we may use the term coalition. (Coalitions are affiliations formed for specific purposes. Affiliations may also come about spontaneously. The distinction is not important in the present context.) In most models certain specific forms of affiliation or coalition are regarded as institutional data. In the most usual formulation of a general equilibrium system, the most important of these specific forms are households, firms and markets (or the market). There are of course other forms of affiliation, such as trade unions, employer's organizations, national and ethnic groupings and even the Marxian or other social classes. However, these (or some of these) make only occasional appearances in special versions of general equilibrium theory. The problem now is to find a more general formulation of an economy in which coalitions are parametric data, that is, though specific forms of coalition exist for shorter or longer periods, the formulation must not be tied to any one specific form as an institutional datum.

First of all we may avail ourselves of another simplification of genius, namely, Carl Menger's idea of goods of lower and higher order. Instead of speaking of consumer goods and factors of production, we now speak only of various orders of economic goods. The first order of economic goods consists of those which may satisfy wants directly, while the higher orders consist of those which may be transformed into first-order goods, given a knowledge of technical constraints.[18] Armed with this we may leave open the question of whether people organize themselves into households and firms. If they do not (if the set of coalitions is empty in this respect) we have a picture of individuals with certain endowments who seek to reach the most preferred position attainable by transforming their endowments either through production or through exchange in the market, the choice depending on whether the technical constraints or the prices in the market offer more favourable terms. Of course, endowments may first be exchanged, then transformed by production, then exchanged again, and so on. Specialization and entrepreneurial activity are thus not excluded from this picture and profit maximization is simply a part of the attempt to reach the most preferred position attainable. In some respects, this very general scheme seems more realistic than the standard view, no doubt influenced by accounting procedures, of firms which neither own nor earn anything in themselves and of households who do nothing but consume. G.S. Becker, for instance, has argued convincingly for the usefulness of regarding households as always faced by the decision either to sell and buy in the market or to produce in the home.[19]

In Menger's scheme the existence of markets is still an institutional datum. We now wish to express all coalitions as parametric data. Fortunately, we can here avail ourselves of the theory of the core of an economy, which economics has borrowed from the theory of games and in which the word coalition is at home.[20] Comparatively minor adjustments have to be made to the picture outlined. There we could visualize a certain allocation of endowments among individuals. The individuals transformed these endowments through production and exchange into the most preferred allocations attainable. For these purposes they formed the coalition of a competitive market which ensured so efficient an exchange of information among individuals that at any one time it was known for how much of any or all of the other goods any one good may be exchanged (that is, there were market prices). The structure of the economy may be such, however, that the exchange of information is less efficient so that only certain opportunities for exchange are known and more than one rate of exchange between any two goods may prevail at any one time. In the extreme case, bilateral bargains are struck between individuals. The theory of the core of an economy is designed to handle such quantity bargaining or barter. The core consists of all those final allocations which, first, are not less preferable to any one individual than the original allocation of endowments and, second, which cannot be blocked by any coalition of a subset of the individuals, where an allocation is blocked if a coalition can come to some arrangement whereby at least one individual reaches a more preferred position and no individual has to accept a less preferable one. The allocations in the core are therefore Pareto optimal and also correspond, *mutatis mutandis*, to those on the contract curve of an exchange economy, introduced by Edgeworth in his 'Mathematical Psychics'.

It is usually shown that as the number of individuals increases under certain conditions, the allocations in the core are reduced in number and in the limit approximate competitive equilibria. For this, no restrictions are usually placed on the kind of coalitions that may be formed, though the cost of forming coalitions may be taken into account. The idea that all conceivable arrangements may be used to form coalitions is somewhat foreign to the present analysis. If instead we regard coalitions as parametric data, so that at any one time certain forms are specified as already existing or known to be feasible (for example, markets of various degrees of perfection, households, firms, trade unions, and so on), the number of Pareto optimal allocations is restricted in a way analogous to the restrictions imposed by the specification of perfectly competitive markets in the Walrasian general equilibrium analysis. Our picture based on Menger's scheme now has to be adjusted only slightly. The choice between production and exchange can no longer be based on the known technical constraints and market prices, but rather on the known technical constraints and the possibilities manifested by the forms of coalition known to exist or to be feasible. It should be noted very clearly that this picture also requires us to regard the technical constraints and the preferences of individuals as parametric and not as casual data.

One further step is required in my generalization of economic systems. Property norms and the attendant laws of contract are known to vary over time and from place to place, just as market forms do. Though individuals may regard specific forms of property norms as institutional data in their daily actions, we shall here form a new data set called property norms and regard it as parametric. One possibility now is that the set of property norms is empty, that is, that there are no property norms. Under these conditions, it is still possible for an individual to have something in his physical possession, and exchange is still possible in the sense that one thing is handed over for another, even though all individuals not involved in the transaction are likely to disregard any of the conditions that may be made.[21] However, the concept of endowments must now be re-examined. Since resources and capital goods cannot be appropriated in accordance with property norms, and one's endowments are restricted to one's energy, strength, skills, intelligence, and so on, and perhaps more importantly to what happens to fall into his hands through the good fortune of being in the right place at the right time. While a person's characteristics, whether they are an endowment or a handicap, may be regarded as parametric data, the person's good fortune is merely something that happens and is therefore a casual datum.

The parametric data that we have now assembled are the following:

1. the various characteristics of individuals;
2. property norms with concomitant laws of contract;
3. coalitions or affiliations among individuals;
4. technical constraints; and
5. the preferences of individuals.

3.5.7 The variability of parametric data

I shall investigate one final question and, needless to say, it is this question that I have been trying to lead up to in this section. The question is whether the five basic data sets listed at the end of section 3.5.6 should be regarded, on the basis of general considerations, as equally variable. All five were classified as parametric, but the durations of their specific forms or values may nevertheless differ greatly. Unfortunately, there is still a great deal to be done before we can actually consider this question.

We shall not be able to go far without running into a very fundamental and problematical distinction in economics. It is that between stock and flow equilibria, or that between goods on the one hand and productive services and consumption on the other. It seems to me that some of the thorniest problems in economic theory ultimately stem from this distinction and that much could still be written on it. I want to stress therefore that I shall here touch on this distinction merely in so far as it pertains to the question

I have set; and for this one does not have to look into it very deeply. I shall argue that stock and flow equilibria correspond to a deterministic and the stationary-state conception of equilibrium respectively. As I indicated in sections 3.5.1 and 3.5.2, these conceptions are not entirely compatible, though Walras managed to graft them together very ingeniously.

In section 3.5.1, we saw that a pure exchange economy can be seen as a pure structure consisting of demand and excess demand functions linked by the equilibrium conditions. I tried to show that a pure structure is an axiomatic construct. However, on having another look at demand functions in exchange economies in section 3.5.5, we saw that they have two constituent elements, initial endowments and preferences, and that at least one of these, the initial endowments, cannot be regarded as parametric data. On this view, then, exchange economies are not pure structures. We may now use the general scheme developed in section 3.5.6 to isolate the parametric or structural data. In the simple exchange economy, these consist of property norms and laws of contract, the coalition or affiliation of a market and the preferences of individuals. There are also the initial endowments which are not parametric, and the analysis now consists of the transformation of the allocation of initial endowments by means of the structure into the equilibrium allocation. In an empirical context, in fact, one can become aware of the parametric or structural data only if some process such as this transformation is taking place. Now, we have seen that a deterministic model focuses attention on an initial and a final state and that the connection between the two may be variously regarded as governing laws, functional relations or a structure of interlinked *ex ante* fact. We may conclude therefore that the equilibrium of an exchange economy is essentially a deterministic conception of equilibrium, in which the allocation of initial endowments constitutes the initial state and the equilibrium allocation the final state. The only difference between the deterministic conception of equilibrium and the kind of deterministic conception of historical processes that I dealt with in Chapter 2, is that equilibrating processes may come to an end, if equilibrium is reached, whereas deterministic historical processes are thought to continue to the end of time. One may regard this difference as a difference of degree depending on the number of *ex ante* facts linked. If the equilibrium allocation were the exogenous datum of a further system which is linked to yet further systems ad infinitum, the equilibrating process would not come to an end.

The significance of recognizing exchange equilibrium as a deterministic conception is the following: The initial and final states of a deterministic model are points in time and the data that go into the model and the results that come out of it therefore pertain to points in time and are necessarily *stock* concepts. In the exchange economy, the initial state is a certain allocation of *goods* (not services) and the final state is the equilibrium allocation of *goods*. Continuous time enters the model only with respect to the process between initial and final states, but this is an equilibrating process and not an equilibrium flow, and in any case it is a process in which comparative statics is not greatly interested, at least not in its entirety. Moreover, as we have seen, once equilibrium has been reached then, in the absence of disturbances from outside the system, the analysis has come to an end. It concerned the reshuffling of stocks, and once they have been reshuffled there is nothing further to be said. We may assume of course that individuals find one allocation preferable to another, because they intend to consume goods or their services, or to use them as capital goods. But consumption and the performance of services must take place outside the analysis: they cannot be the subject of a deterministic equilibrium analysis, for it is designed to handle changes in stocks but not to handle equilibrium flows.

Equilibrium flows of services and their consumption belong to the stationary-state conception of equilibrium. We have seen that when the deterministic initial and final states are tied up in an additional relation, a closed system of interlinked relations is formed. Without initial and final states, there is no reference to points of time and therefore no reference to stock concepts.

To my mind the best interpretation of an axiomatic construct is that it does not refer to time at all. However, as I pointed out in section 3.5.2, many economists have taken it to refer to continuous time, to flow concepts and to a stationary-state conception of equilibrium. The simplest stationary state requires as data only the parametric preference of individuals, the abilities and aptitudes of individuals for performing services and some coalition of affiliation that brings individuals together repeatedly so that they may *perform their services* for each other. There is not even a need for a market equipped with the Walrasian price crier, except that the price crier must be seen standing in the background as the man responsible for the bringing about the stationary-state at some time in the past.

One can introduce stock concepts into a stationary state, and specify certain property norms, without doing too much harm to the analysis. One can of course quite easily visualize a flow of owned goods. One can also deal with a situation where the flow of goods consumed is less than that produced, for accumulation can be handled, with appropriate additional assumptions, by some dynamic version of the stationary state – the uniformly progressive economy, steady state growth or a golden age. Even if the individuals in the stationary state are not content merely to provide services and to own and accumulate goods that provide services, but actually want to exchange the goods they own, then no great harm need be done, for one can take a macro-economic point of view to which the stationary state lends itself, and say that it does not matter which individual owns what as long as the aggregate flows remain the same.[22] None of this alters the fact that the stationary state is really concerned with equilibrium flows of services and their consumption and that the stocks in it are no more than visual aids. Stocks are not the subject of the analysis, as they are in the deterministic conception of individuals who hold certain stocks but would prefer, and are able, to hold different stocks.

Menger's analysis is entirely in terms of goods, and in so far it is compatible with the deterministic conception. (We shall see in Chapter 5 that it is deterministic only in a very restricted sense.) Walras, however, combined the two conceptions most ingeniously. From lessons 5 to 16 of the *Elements*, in which he deals with the exchange of first two and then several commodities for one another, he uses the deterministic conception of equilibrium. Then he comes to his theory of production.[23] He first distinguishes between capital and income. Capital consists of goods (or social wealth) which may be used repeatedly, and income of those that may be used only once. Both are goods and may be exchanged for one another. But since capital can be used repeatedly, 'the flow of uses evidently constitutes a flow of income' (p 213). He then distinguishes between landed, human and produced capital, each of which in turn may provide either consumers' services or productive services. He also lists various categories of inventories which provide a 'service d'approvisionnement'. There are now two distinct markets, one for products and one for services (pp 222ff). The entrepreneur appears and sells products and buys services while everybody else buys products and sells and buys services. Though distinct, the markets are made comparable because services are not actually traded on the services market, but rather service contracts or promises to deliver against goods in the future and such contracts or promises can be traded like goods. Services are thus transformed into stock concepts and the whole analysis can be carried out in terms of a deterministic conception. Soon the terms for goods and services become almost interchangeable. 'Services, then, are also commodities the utility of which can be expressed for each individual by a want or utility equation' (p 237).

However, Walras also keeps a foot in the stationary state. When both markets are in equilibrium and the prices of goods are equal to the cost of the services that went into them 'we may even go so far as to abstract from entrepreneurs and simply consider the productive services as being, in a certain sense, exchanged directly for one another, instead of being exchanged first against products, and then against productive services' (p 225). The stationary state comes into its own in his theory of circulation and money. 'Once equilibrium has been achieved ... the actual transfer of services will begin immediately and will continue in a given manner during the whole period of time considered' (p 316). But now stock concepts are accounted for by the 'service d'approvisionnement' of inventories and cash balances (pp 319f). Goods have been transformed into services of various kinds. No doubt Walras was not the first to combine the deterministic and stationary-state conceptions of equilibrium, but his ingenuity has surely contributed much to the integrated picture presented by economic theory today. It has also left economists with some peculiar puzzles, not least in capital theory, which has to straddle both conceptions.

Now, I have gone into all this because it will be very important for my purposes to keep the two conceptions of equilibrium apart. I have been and shall be concerned with structure and random change in economics, and the stationary-state conception of equilibrium is not adapted to analysing the effects of random changes. We saw in section 3.5.2 that change in a stationary state must be seen as an adjustment in a parameter. When such changes are frequent, flow equilibrium becomes meaningless to an observer and the course of events becomes indistinguishable from one for which there is no theoretical understanding. In other words, when there are frequent random changes, the stationary state ceases to be interesting in itself and attention must be focused on the transition from one stationary state to another. The stationary-state conception, however, cannot give a visual meaning to such a transition and recourse must be had to the deterministic conception. I have suggested that the stationary state is a visual representation of an axiomatic construct, and it seems to be that no one has yet been able to extend this representation to the case where we simply change the axioms or assumptions. For this, economic theory has resorted to equilibrating processes (as distinct from equilibrium flows) and these are in the sphere of the deterministic conception.[24] Though theory has not been able to cope adequately with the problem of false trading, equilibrating processes at least have a visual, and therefore in principle an empirical, meaning. Since I want to avoid the problem of dealing with stocks and flows at the same time (or is it period), which in my opinion is a problem of dealing with a mixture of incompatible conceptions, I shall confine my remarks to the deterministic conception of equilibrium. Towards the end of the study, I shall suggest a variant conception in which *ex post* and *ex ante* facts take the place of deterministic notions.

I shall use the adaption of Menger's scheme, which I developed in section 3.5.6, as a vehicle for examining the deterministic conception. It is a very general scheme and, when the parametric data are appropriately specified as institutional data, may be applied to the analysis of subsistence economy, a market economy, a so-called mixed economy and even a strictly collective socialist economy. Unless otherwise specified, I shall use it in the context of an economy with fairly competitive markets, and property norms which would be considered normal in a free-enterprise economy. There are purposefully acting individuals and there may be certain coalitions of individuals, such as households and firms, which are purposefully acting units. At any one time individuals and/or coalitions have certain holdings of goods. The analysis concerns the transformation of these holdings of goods into the most preferred holdings attainable either by production according to known technical constraints or by exchange according to the institutions for exchange that may exist. Flows of services and consumption do not enter the analysis as such and production is therefore not a flow of services but a transformation of goods.

A good may be described more or less as Menger described an economic good,[25] that is, in short, something which its holder believes capable of satisfying a part of his needs or requirements (*Bedürfnisse*) or something that may be converted into such a thing, and of which there is less than is required. A good may be either tangible or intangible, alienable or inalienable and capable of being used in production or not, but in order to enter the analysis it must be either alienable or capable of being used in production or both. A capital good is a good that is used in a production transformation without being transformed itself and is thus not distinguishable by any physical properties but by the way it is used in a particular production transformation. The first of the five basic data sets I listed at the end of section 3.5.6, namely, the abilities and aptitudes of individuals, may enter the analysis in two ways. First, it appears from the definition of a capital good that a person must regard himself as an inalienable capital good, and I take it that this is what Walras meant by what is personal capital in Jaffe's translation. (A person's abilities may of course affect not only his technical constraints but also his opportunities for exchange, as does a salesman's persuasiveness in an imperfect market. But this merely underlines the similarity between production and exchange in this scheme.) Second, there is the other device used by Walras, namely, that among the goods a person may hold are the promises to render services in the future, though this requires that others should also regard these promises as goods. Only the promises enter the analysis; not the actual performance of the services.

There are some very conspicuous difficulties in this scheme. If individuals trade in promises, and not only to render services but also to cede the use of goods or to repay debts in the future, then surely one cannot exclude the expectations and plans of individuals from the analysis, even if one can exclude the actual carrying out of the promises. This may be granted, but it is not an insuperable difficulty for one may with a fairly good conscience include expectations and plans in the preference of individuals. After all, one cannot expect to get far in the analysis of human affairs if by preferences one understands only that an individual can decide between bundles of goods such as deciding between tea and coffee or apple and bananas. What one expects to happen and what one plans to do must have something to do with one's choice. However, there are related difficulties that are not so easily put aside. One may ask how the most preferred holding of goods attainable is affected not by events after the final state of the analysis, but by events between the initial and final states. Does the individual take into account what Jevons called the disutility of work in deciding what is attainable? Also, the individual can obviously have more in the final state if the individual does not consume in the meantime. Can consumption really be left out of account then? Again, if production transformations take time they have an opportunity cost in the form of services that could have been sold. Can services be left out of account then? These difficulties, however, are also inherent in standard neo-classical theory. They are not noticeable if we think in terms of instantaneous adjustments from one flow equilibrium to another.[26]

We are at last ready to consider the question I have set, namely, whether the five basic data sets listed at the end of section 3.5.6 should be regarded as equally variable. Let us note that the initial state at which neo-classical analysis starts is any state and that no questions are asked about its origins.[27] A deterministic approach to equilibrium must of course start at such a state, for if it were traced back to some prior state, the initial state would simply be pushed back further in time. Let us also note that if stock equilibrium is reached then the deterministic part of the analysis, which is all that we are interested in here, comes to an end. However, if at this stage a change occurs about the origin of which we know nothing (that is, one that we cannot trace back to some prior state), and if as a result the holdings of goods are no longer equilibrium holdings, then there is a new initial state and the analysis starts all over again. If now the frequency of such random changes increases to such an extent that there is always a change just as equilibrium is reached, then the analysis need never stop and we would be concerned with continuous time, but not with equilibrium flows. We would also be concerned with comparative statics and not with a dynamic model. Of course, we could also say that the rate of random changes is such that the successive equilibria are never reached, but then we must concern ourselves with the dynamic process between initial and final states, but still not with equilibrium flows.

One may now ask what kind of change may disturb a stock equilibrium, that is, change an equilibrium allocation of goods into an initial state. The kind of change is to be identified not by what prior state has brought about a change but simply by what it is that changes. First, an equilibrium holding of goods may be changed into a disequilibrium holding without a change in any of the parametric data. This occurs whenever a stroke of bad luck or good fortune happens to an individual. I pointed out at the end of section 3.5.6 that there is always a trace of this in the income concept. Lightning may destroy a house and its owner may then not like the goods owned, which then includes a burnt-out house and possibly a large bank balance. Economic theory has shown adequately that one's efforts to reach a new equilibrium holding of goods may affect the equilibrium holdings of many or all other individuals. However, I am more interested in changes in the parametric data. Being parametric, any of them may change.

A change in the technical constraints or in the prevalent forms of coalition may make more preferred stock holdings attainable and so create disequilibrium. A change in an individual's preference ordering, though it leaves unaffected what is actually held and what is attainable, nevertheless makes the individual's holding of goods a disequilibrium holding. A change in only one individual's preference ordering may make all other individuals' holdings into disequilibrium holdings. It appears that a change in any parameter may create disequilibrium and therefore a new initial state. All the parametric data may of course take on different specific forms, but if they all change equally rapidly we have again the flux of Heraclitus and no deterministic model can help us to understand the course of events.

Let us suppose, purely for heuristic purposes, that the preference orderings of individuals change far more rapidly than any of the other parametric data. Whenever a change occurs an equilibrating movement starts and it takes on a definite direction because the other parametric data have not changed. As the frequency of changes in preference orderings increases, so we approach a continuously changing course of events. But as we make this approach, the changes which we observe are not entirely random because they are still constrained by the parameters that do not change.[28] The picture we get in this way has some considerable advantages. First, there is the continually changing scene which is so uncharacteristic of the stationary state but so very characteristic of our everyday experience. Second, there are parametric data with constant forms which are not unfamiliar to us in our everyday experience. These ought to give a truncated determinism much scope for finding 'qualitative restrictions' of the Samuelson type. Third, there are preferences changing continually, not to prove too offensive to the free-will camp of economics.

With these apparent advantages, the picture, I think, is worth further investigation. I argued in section 3.5.3 that if comparative statics is not to be an exercise in empty logic, one has to commit oneself to an empirical judgement about the relative variability of data. Accordingly, I shall devote the remaining part of this study to some general consideration that may be relevant in an assessment of the variability of preference orderings. In the main, my approach will be to apply the conceptual framework developed in Chapter 2 to the question of the variability of preferences orderings. In that conceptual framework, it will be remembered, one does not start with a notion or working hypothesis such as that every human being has a free and independent will or a consistent preference ordering. Rather one considers how an observer becomes aware of something, in this case of the will or preferences of people *other than the observer*. The observer's introspection may reveal preferences and perhaps even how changeable they are. But it cannot tell the observer what others prefer most and how often they change their minds. The question that I shall therefore pursue, in some instances explicitly and in others implicitly, is whether the preferences of other people are to an observer *ex ante* or *ex post* facts.

The question of the variability of preference orderings really involves two entities, namely, the individual's preference ordering itself and the so-called law of demand which, as Hicks put it, 'remains what it always was, the centre of the whole matter'.[29] The preference hypothesis is of course an attempt to account for the law of demand, and the latter is then seen to arise out of certain common properties of the individual preference orderings. However, it may well turn out that it is difficult to accord the two the same empirical status and they therefore have to be kept apart. There is yet another matter that has to be treated separately, namely, the question of maximizing behaviour, for it is not a question of how long preference orderings last or of whether they have any common properties, but rather whether we should talk about preferences at all. It is not a question that can be decided on empirical grounds. However, that is the subject of the next chapter.

Notes

[1] I shall, however, not attempt to do justice to the considerable sophistication of modern general equilibrium theory or to survey the extensive literature on it; nor would I be able to do so. This field of economic theory seems to have been elaborated so much that anyone who sets about making himself expert in it, could probably do little else. The writer on the philosophy of science who deals with fields as diverse as, say, nuclear physics, botany and sociology, may seem to lay himself open to a charge of dilettantism, but his task is to see whether there are any links between subjects that would make more general formulations possible. I think a case can be made out for a similar service within the field of economics itself. We all know that it is no longer feasible to know much about what one's colleagues are doing, but even the most eminent economists in not very widely differing fields find the exchange of ideas difficult. A good example of this may be found in the inaugural lecture of Professor Hahn of Cambridge (*On the Notion of Equilibrium in Economics* [Cambridge: Cambridge University Press, 1973]). He says on page 6, 'Professor Kaldor on hearing what I proposed to discuss on this occasion urged me to take notice of his latest paper in the "Economic Journal"' ('The Irrelevance of Equilibrium Economics', 1972). The academic cross-pollination quickly proves sterile. On page 8 he says: 'Professor Kaldor's theory of what it is that Debreu's book might be about is thus incorrect', and he goes on to say in more, but not much more, polite terms that Professor Kaldor simply does not know what the mathematical general equilibrium economists have been doing.

[2] Schumpeter thought that the term was coined in 1916 by Franz Oppenheimer, whom he calls a 'positivist' sociologist. See *History of Economic Analysis* (London: Allen & Unwin, 1954), pp 855 and 965.

[3] Alfred Marshall, *Principles of Economics* (8th edn, London: Macmillan, 1920) p 468. The illustration has of course become a favourite of textbooks since then. Marshall used it as a step in the reasoning by which he compares the magnitudes of the gross receipts from a tax and the loss of consumers' surplus.

4   For a discussion of the relation between static and dynamic models, see Robert Kuenne, *The Theory of General Economic Equilibrium* (Princeton: Princeton University Press, 1963) pp 13–15 and 455–62. See also pp 27–31 for his discussion of the Walrasian and Marshallian stability conditions.

5   Leon Walras, *Elements of Pure Economics*, translated from the edition of 1926 by William Jaffé (London: Allen & Unwin, 1954) p 71. 'This much is certain, however, that the physico-mathematical sciences [Walras classifies economics as such], like the mathematical sciences, in the narrow sense, do go beyond experience as soon as they have drawn the type concepts from it. From real-type concepts, these sciences abstract ideal-type concepts which they define, and then, on the basis of these definitions they construct a priori the whole framework of their theorems and proofs.' (See also pp 83–6 where Walras describes how he sees markets.) His ideal type would be described, I think, as an Aristotelean conception; it is still largely current in economics today. If instead one accepts the Kantian view that empirical knowledge is formed, so to say, by a fusion of *a priori* categories and external stimuli, then at least the logical categories of mathematics (for example, number, set) are seen to be more intimately related to that which we can visualize. This does not mean, however, that we can visualize a purely a priori axiomatic construct like a five-dimensional space.

6   For an excellent discussion of these issues see M.R. Cohen and E. Nagel, *An Introduction to Logic and Scientific Methods* (London: Routledge, 1934) pp 129–50. See also I.M.D. Little, *A Critique of Welfare Economics* 2nd edn (Oxford: Clarendon, 1957) pp 19f.

7   K.J. Arrow and F.H. Hahn, *General Competitive Analysis* (San Francisco and Edinburgh: Holden-Day, Oliver & Boyd, 1971). The quotations come from p 245 and p 12.

8   *Ibid.*, p 278.

9   F.H. Hahn, *On the Notion of Equilibrium in Economics* (Cambridge: Cambridge University Press, 1973). Page numbers in the text refer to this publication.

10 Walras, *op. cit.* (note 5) pp 84ff.

11 The idea of the stationary state is very old. Schumpeter cites Plato's 'Republic' as an example. It is to be seen in Quesnay's system, and with Ricardo and the classical school in general it came to play quite dominant role. (See Schumpeter, *op. cit.* [note 2] pp 55–5, 562–3, 964–7.) The classical school saw the stationary state as a state of affairs towards which actual economies were heading, and in it not only production and prices but also, for example, population would remain unchanged. In those cases in which static equilibrium is regarded as a newer version of the stationary state, it is distinguished from the older version by the fact that it is phrased in terms of preferences and is regarded purely as an analytical tool. Marshall, *op. cit.* (note 3) pp 366–9, discusses the stationary state without much enthusiasm, adding his variant of the representative firm. However, he says that 'less violent assumptions' are made for 'statical method' and that 'we suppose it for the time to be reduced to a stationary state'. A.C. Pigou, *The Economics of Stationary States* (London: Macmillan, 1935) examines the question in great detail. J.R. Hicks, *Value and Capital* (Oxford: Clarendon, 1939) says on page 58 that a market is in static equilibrium 'if every person is acting in such a way as to reach his most preferred position, subject to the opportunities open to him'. On pages 116 to 119 he discusses static equilibria in relation to stationary states and in the final paragraph of the book rejects the latter in favour of a more whole-heartedly deterministic dynamic approach. F. Zeuthen, *Economic Theory and Method* (London: Longmans, 1955) pp 33–4 discusses the issue and evinces misgivings about visualizing static equilibrium at all. J.E. Meade, *The Stationary Economy* (London: Allen & Unwin, 1965) p 184 describes 'stationary equilibrium' as a position where there is 'no incentive on the part of any buyer or seller to change any price or any quantity of anything which is being bought or sold'. One may also regard F.H. Hahn's

notion of an equilibrium that reflects the sequential character of economies, which I consider in section 3.3 of this chapter, as a version of the stationary state. Note that in the way I have put it in the text, static equilibrium is merely a Pareto optimal position which becomes a competitive equilibrium if markets are one of the institutional constraints.

[12] Milton Friedman, 'The Marshallian Demand Curve', *The Journal of Political Economy*, LVII, 1949, p 464.

[13] P.A. Samuelson, *Economics*, 9th edn (New York: McGraw-Hill, 1973), pp 59 and 429.

[14] Friedman, *op. cit.* (note 12) pp 481–2.

[15] See J.R. Hicks, *Value and Capital* (Oxford: Clarendon, 1939) pp 32 and 128–9 and Marshall, *op. cit.* (note 3) p 335. Marshall also supported his argument for the constant purchasing power of money by referring to dealers whose purchases do not diminish their resources since they look to re-selling. Hicks does not mention this, presumably because markets in intermediate goods have not been stressed in more recent theory. As I shall indicate in note 16, the matter may be of some importance.

[16] See Friedman, *op. cit.* (note 12) pp 465–6, 475 and 484. With regard to these rival interpretations, it may be worth noting the role Marshall ascribed to dealers which I mentioned in note 15 above. On page 335 of the 'Principles', Marshall defends the assumption of a constant purchasing power of money by saying: 'There may indeed be individuals of whom this is not true, but there are sure to be present some dealers with large stocks of money at their command; and their influence steadies the market.'

[17] Walras, *op. cit.* (note 5) Lessons 6, 7, 9, 11 and 12. For an example of the manna-gatherer economy, see C.E. Ferguson, *Microeconomic Theory*, revised edn (Homewood: Richard D. Irwin, 1969) pp 422ff.

[18] Menger sets out the idea in sections 2 and 3 of Chapter 1 of his *Grundsätze der Volkswirthschaftslehre* of 1871. See either *The Collected Works of Carl Menger*, vol 1 (London: London School of Economics Reprints, 1934), or *Principles of Economics*, translated and edited by J. Dingwall and B. Hoselitz (Glencoe: The Free Press, 1950). Schumpeter, *op. cit.* (note 2) p 913, speaks of Menger's idea as an 'analytic device that looks so simple or even trite and was nevertheless a genuine stroke of genius'. On the other hand, in the introduction to the English translation mentioned above, F.H. Knight says (on p 25): 'Perhaps the most serious defect in Menger's economic system is his view of production as a process of converting goods of higher order into goods of lower order.' One may presume that Böhm-Bawerk drew on Menger's idea for his concept of capital. Walras, *op. cit.* (note 5) p 73, speaks of indirect and direct utilities, and says that one of the aims of industry is to transform the former into the latter. He does not make further use of the idea.

[19] Gary S. Becker, 'A Theory of the Allocation of Time', *Economic Journal*, 75, 1965, pp 493–517. Since many of the chores done around the home can often be contracted out to firms, households can hardly avoid this decision. Firms of course also have to decide whether to buy components or to make them themselves. Becker discusses the allocation of time within the institutional setting of firms and households, and therefore does not seek the level of generality of Menger's analysis. He cannot of course distinguish between firms and households on the basis of where production decisions are made. Instead the distinction is based on the discretionary control over time. Individuals have discretionary control over the allocation of time between production and consumption in households, but not in firms (p 496). He later admits that it is difficult to make this distinction in a long-run view (p 504). Menger, on the other hand, introduces firms only in his second last chapter where he speaks of commodities (*Waren*) as economic goods made specifically for sale. He treats the firm as a particular historical development which must be seen as a special case in his more general scheme.

[20] For a short explanation of the core of an economy see Arrow and Hahn, *op. cit.* (note 7) pp 183–7 and 198.

[21] See John R. Commons, *Institutional Economics* (Madison: University of Wisconsin Press, 1961; first published Macmillan, 1934). Commons laid great stress on the distinction between exchange, which can be carried out between two individuals, and a transaction, which requires the consent of other individuals. The one is 'physical delivery of physical control over commodities or metallic money' and the other 'legal transfer of legal control' (p 60). 'Transactions ... are not the "exchange of commodities" in the physical sense of "delivery", they are the alienation and acquisition, between individuals, of the rights of future ownership of physical things as determined by the collective working rules of society' (p 58). 'The individual does not transfer ownership. Only the state ... by operation of law as interpreted by the courts, transfers ownership by reading intentions into the minds of participants in a transaction' (p 60). To Commons, generally recognized as having been one of the three leading figures in American Institutionalism, all this was of great significance, because he maintained that orthodox economics did not make the distinction and that it gave a double meaning to wealth, namely, 'the physical meaning of holding the materials of nature for one's own use in production or consumption, and the proprietary meaning ... namely the right to exclude others and to withheld from them what they want but do not own' (p 302). (See also Chapter 1, footnote 4.) Orthodox economists, however, did not stress the proprietary meaning 'and they therefore concealed the field of institutional economics' and it was 'this concealed ownership side of the double meaning of Wealth that angered the heterodox economists' among whom was Marx (p 55). Statute law, ethics, customs and judicial decision, 'all these might be eliminated by assuming that ownership was identical with the materials owned, in order to construct a theory of pure economics based solely on the physical exchange of materials and services' (p 56).

[22] Of course, the aggregate flows may not remain the same if prices of goods change. See John Hicks, *Capital and Growth* (Oxford: Clarendon, 1965) especially pp 131–69 and 194–7. Hicks examines the question of prices in growth equilibrium at length, and 'the stationary state is a growth equilibrium with a growth rate zero' (p 133). He says that 'if we really stick to aggregates then (as has been shown) we are committing ourselves to a Fixprice theory – meaning by that ... a theory in which prices only change exogenously, not as consequence of changes in other aspects of the system' (p 183). As I have stressed, I am here touching on the question of stocks and flows for a particular purpose only. It is obviously a vast subject.

[23] Walras, *op. cit.* (note 5) pp 211ff.

[24] In this connection it is interesting to note that in what Hicks calls the temporary equilibrium method, which for instance Lindahl and he has used, the equilibrating is confined to the beginning of a period, the Monday trading day, while equilibrium flows occur in the rest of the period. Hicks's discussion of the 'Traverse' also concerns equilibrating processes in a context of equilibrium flows. See Hicks, *op. cit.* (note 15) chapters IX–XXIL, especially pp 115–40, and *op. cit.* (note 22) pp 58–83 and 183–97. In almost the whole of the latter of these works, Hicks grapples with the question of stocks and flows. Chapter VII is explicitly devoted to the question. Hicks's discussion, however, relates to dynamic models and I have explicitly excluded these from my discussion not only to keep the study within manageable proportions (and this is an important reason) but also because I can reach the point I want to make with comparative statics only.

[25] Menger, *op. cit.* (note 18) pp 1–76 and 225–49, English translation, pp 51–113 and 236–56.

[26] The temporary equilibrium method also separates the deterministic from the stationary-state conceptions of equilibrium, that is, the equilibrating from the equilibrium processes. (See note 24.) Consumption, work and other services are suspended on the Monday trading day while the flows for the rest of the week are decided upon through a price

mechanism. While the flows occur, all trading is suspended. If the Monday trading were to establish a final instead of a temporary equilibrium, the activity in the rest of the week would hardly be of any interest. The promises exchanged and the production decisions made on the Monday would be all that mattered. If trading activities, consumption and production take place over the same period, the problems mentioned in the text arise. Walras had already effected the separation. At the end of his discussion of recontracting by 'tickets' (*op. cit.* [note 5] p 242), he said: 'Thus, equilibrium in production will first be established *in principle*. Then it will be established *effectively* through the reciprocal exchange between services employed and products manufactured *within a given period of time during which no change in the data is allowed*' (his italics). He did not say what those parametric data actually was.

[27] For instance, Walras started even his analysis of the equilibrium of circulation with random endowments (*op. cit.* [note 5] p 318). He said: 'Thus, we shall imagine an economy establishing this equilibrium above over a given period of time during which no changes take place in the data of the problem. We shall, accordingly endow our bank-owners, workers and capitalists, viewed as consumers with random quantities of circulating capital and money, just as we endowed them before with random quantities of fixed assets in the form of landed capital, personal capital and capital proper.' Walras did not make it clear whether all the initial endowments were included in the data that were not to change.

[28] One could here resort to the device I used earlier of representing a deterministic model by ordinary functional equations. For distance, let $y_1 = f_1(y_0)$, $y_2 = f_2(y_1)$ … $y_n = f_n(y_{n-1})$ and f_1 contain the set of parameters a_1, b, c…m, f_2 the set a_2, b, c…m, f_n the set a_n, b, …m with b, c,…m retaining the same values in each case. One can then start with any y_0 and give a its values from 1 to n. The resulting values of y from 1 to n would be constrained by the constants b, c,…m. It would be much easier to say that $y=g(x)$ and $x=h(a)$ where g contains the parameters b, c…m. But if the parameter a relating to preferences were taken out of the deterministic model, the model would lose its economic meaning, that is, $y=g(x)$ would have no economic meaning if g related only to technical constraints, coalitions, property norms, and so on, and x to initial endowments. In order to visualize a process here one requires the notion of purposeful action.

[29] J.R. Hicks, *A Revision of Demand Theory* (Oxford: Clarendon, 1956) p 59.

4

Rational Action

4.1 Maximizing behaviour

In *The Theory of Political Economy*, Jevons made the following remark on the logical method of economics:

> I think that John Stuart Mill is substantially correct considering our science to be a case of what he calls the Physical or Concrete Deductive Method; he considers that we may start from some obvious psychological law, as for instance, that a greater gain is preferred to a smaller one, and we may then reason downwards, and predict the phenomena which will be produced in society by such law.[1]

Let us take this 'obvious psychological law' as one way of expressing the assumption of maximizing behaviour. The question that I want to consider in this chapter is whether this law, or other versions of it, can be said to have a useful meaning when put in categorical terms such as Jevons used. As it stands in this quotation, it certainly seems that it is neither a law nor psychological, though it may be obvious. It is simply a statement that puts the words 'gain' and 'prefer' into their logical relation to each other, so that if one knows the meaning of one of them one may know the meaning of the other. However, there are at least two ways of understanding the statement in which it is not a tautology. It may be taken as generally understood what kind of situation or holding of goods people would regard as a gain over another one, or, even if this is not known, it may be useful in the analysis of human affairs to say that people do prefer some things to others. The former implies that we have definite empirical knowledge about people; the latter also is a definite empirical assertion about people since we would not normally say, for instance, that stones have preferences.

It appears from the context of Jevons's discussion that he had the former of these in mind. He went on to say that he did not regard Mill's concrete deductive method as *a* method but as *the* inductive method, by which he understood more or less what we would now call the falsifiability criterion. He suggested that 'psychological laws', such as the one in question here, could be tested by deducing particular situations from them which could be checked by observation. It is hard to see how one could reason this from the general to the particular without prior knowledge of what people regard as smaller and greater gains. In the so-called classical political economy there was in fact the implicit assumption that one did have such prior knowledge. Mill's actual words, to which Jevons referred, are the following: 'There is, for example, one large class of social phenomena, in which the immediately determining causes are principally those which act through the desire of wealth; and in which the psychological law mainly concerned is the familiar one, that a greater gain is preferred to a smaller.'[2] It was taken as understood that everyone knew what was meant by wealth or material welfare. Despite much circumlocution, wealth simply meant assets (valued at prices which had to be explained without a coherent theory of demand); and in the layperson's turn of phrase, more money is preferred to less.[3]

English political economy, as compiled and interpreted by Mill, therefore did not concern itself with preferences in general, but with one particular preference, that for more wealth rather than less, and this preference was taken to be common to everyone.[4] The word 'preference' is of course unnecessary when one is not dealing with choice between goods differing qualitatively, but rather with a desire for greater quantities of the same thing. One may then speak of maximizing behaviour. Before the advent of the 'marginal revolution' and of the application of calculus to economics, maximizing behaviour simply meant that the economic individual tried to get the most wealth for the least effort, and outlay of wealth already in its possession. When political economists came under attack, notably from Carlyle and Ruskin, for holding a very one-sided view of human nature, they were at pains to point out that they did not consider the economic person to be the whole person. Mill, for instance, said:

> There is, perhaps, no action of a man's life in which he is neither under the immediate nor under the remote influence of any impulse but the mere desire of wealth. With respect to those parts of human conduct of which wealth is not even the principal object, to these political economy does not pretend that its conclusions are applicable. But there are also certain departments of human affairs, in which the acquisition of wealth is the main and acknowledged end.[5]

When political economists were attacked on quite different grounds, namely that their analysis depended upon specific institutions found in a particular historical setting, their arguments often took on a different direction. We may turn again to Mill. He admitted that political economists had attempted 'to construct a permanent fabric out of transitory materials' and had taken for granted 'the immutability of arrangements of society'. More specifically, the Ricardian theory of distribution in terms of rent, profits and wages was not really applicable outside England and Scotland where the requisite institutions were found. However, he argued that this did not mean that the scope of the subject was extremely limited. 'Though many of its conclusions are only locally true, its method of investigation is applicable universally.'[6] In other words, the 'economic principle' was applicable in all kinds of institutional settings, where presumably different criteria for what is to be regarded as wealth also applied. The ground was being shifted from an assertion of what, among other things, people prefer, to an assertion that they do have preferences. These, as I have already indicated, are two ways of interpreting maximizing behaviour and I hope to show that they are really quite different. When demand theory was incorporated in the mainstream of economics after the 1870s, the distinction became more important. The assumption of a desire for wealth could not in itself provide political economy with a theory of demand. When people have to choose between various asset holdings, they may well always choose the one with the highest value, but this tells one nothing about how they will choose between heterogeneous consumer goods. Demand had not been neglected altogether. It is well-known that Mill had developed a form of demand and supply analysis, but it had remained ancillary to the main principles. In order to bring demand into the body of a coherent theory, the marginal and subjective approaches had to consider a far wider range of preferences. Production and distribution, as they eventually emerged in neo-classical theory, were still based on the old assumption. Entrepreneurs maximized profits; factors of production were hired out at the highest rates of remuneration available (with due allowance for the unpleasantness of work in the case of labour services). But the question of how maximized incomes would be used in the purchase of consumer goods called for a more comprehensive view of preferences. Something other than wealth had to be maximized.

To Jevons the idea that consumers maximize utility or satisfaction must have seemed a logical application of Mill's 'obvious psychological law'. He appears to have been much impressed by Bentham's felicific calculus, that is, the estimation of quantities of pleasure and pain, and even described his theory as 'the mechanics of utility and self-interest' which was 'as self-evident as are the elements of Euclid'.[7] He was not, of course, the first to see satisfaction in such quantitative terms. A long line of utility theorists on the Continent had been, in Schumpeter's words, 'Benthamites by anticipation'.[8] However, for those who are not so convinced by the calculus of pleasure and pain, there is an obvious difference between the maximizing of wealth, profits and income and the maximizing of satisfaction. There is agreement on how we may distinguish empirically between more and less wealth, profit and income, whereas such criteria are not available in the case of utility or satisfaction. There are therefore some means, though in practice they may be difficult, of deciding whether the proposition that firms maximize profits is true or false, as is shown by the criticisms that were levelled against political economy on this matter and are still being levelled against economics today.[9] But how would one decide whether an individual really maximizes his satisfaction? If satisfaction is given the narrowly hedonistic meaning of sensual titillation, the question may be decided, but few have ever understood utility in such a narrow sense.[10] Even Bentham spoke of the pleasures of benevolence and of sympathy. If utility, satisfaction and self-interest are understood broadly enough, then anything that an individual chooses or does may be consistent with maximizing behaviour. 'For if we go very far beyond the grossest gratification of the simplest appetites', Schumpeter observed, 'we come dangerously near to identifying expectation of "pleasure" with all possible motives whatsoever, even with intentional suffering of pain, and then, of course, the doctrine (of self-interest) becomes an empty tautology'.[11] But it is not altogether a tautology, as I have indicated, for when we ascribe self-interest to people, even in the broadest sense, we are still setting them apart from all the things which we do not believe to have self-interest. All this may be quite obvious, but it is important in the present context. In the case of profit maximization, we are specifying what people actually prefer, whereas in the case of utility maximization we are merely saying that people have preferences. The difference between these two interpretations of maximizing behaviour is shown more clearly when the assumption of cardinal utility is dropped, or when we speak only of preference orderings without any overt reference to satisfaction. For the ordinary indifference map we assume that a consumer always prefers a collection A to a collection B if A contains at least as much of each of the two goods as B and more of at least one of the goods. For the revealed preference approach, as Hicks has pointed out, a similar assumption has to be made if the analysis is to serve as an explanation of the propositions of demand theory.[12] More generally, we may say that a collection of goods A is preferred to all other collections if the quantity of each and every good is at its attainable maximum in A, even though the quantities of some but not all goods may also be at their respective maxima in other collections. It may be seen that the old assumption that more is preferred to less is here applied to each good separately and not to one entity such as profits or satisfaction. Obviously, separate maximization may not be possible since there may not be a collection of goods in which the quantity of each and every good is at its attainable maximum.

In order to keep the matter simple, let us consider an individual – and assume it to be a man – who has to choose between two collections of two goods; no other collections are possible within the constraints of his budget. We label the collection A and B and assume that he chooses A. If A is such that it contains more of both goods than B, or more of one good and the same amount of the other, then the individual's choice is consistent with separate maximization. His choice could have been explained even with the principles of the political economy of Mill. But if A contains more of one good and less of the other, the individual's choice cannot be explained in this way. In the absence of some other rule (such as that he always maximizes the quantity of x without regard to the quantity of y, if x and y are the two goods) all we can say is that he chooses A rather than B because he prefers A to B. But what does this tell us? Given the meanings of the words choose and prefer, he could hardly choose otherwise. To say that he chooses A because he prefers B would be nonsense. By ascribing a preference to the individual, we can mean only that we expect there to be collections of goods such as A and B between which the individual is not indifferent. More generally we may say that we do not expect an individual to be indifferent to the various situations open to him, and since he always finds himself in a present situation, we are implying that we expect him often to want to change his present situation into another one, or to prevent it from being changed into another one. In other words, when we ascribe preferences to an individual without stipulating what those preferences are, we can mean only that we expect that individual to act purposefully.[13]

4.2 Consistent preferences

It has long been recognized that economics is concerned with purposeful or rational action. My aim in section 4.1 was not to make this point again, but rather to show that preferences, which are involved in the interpretation of rational action in economics, may be understood either in the sense of their content or in the sense of their form. In this section I want to show how this distinction between the content and the form of preferences has also influenced the interpretation of rational action in economics.

What is the most usual interpretation of rational action in economics? In an article entitled 'Irrational Behavior and Economic Theory', G.S. Becker made the following observation: 'As economic theory became more clearly and precisely formulated, controversy over the meaning of the assumptions diminished greatly, and now everyone more or less agrees that rational behavior simply implies consistent maximization of a well-ordered function, such as a utility or profit function.'[14] Becker went on to argue that rational behaviour thus defined is not necessary for deriving some of the most important theorems of economics, such as the law of demand. Nevertheless, behaviour which does not conform to this definition he called irrational. The crucial words in the definition are 'consistent' and 'well-ordered'. By consistent Becker meant 'that any collection A always gives more, less, or the same utility as any other collection B'. By well-ordered he apparently meant transitive, namely, 'that if A is preferred to B, and B to C, A must be preferred to C'.

Let us first consider the usage of the terms consistent and transitive because it does not seem to be quite uniform. Meade, for instance, speaks of given and consistent preferences and defines consistent to mean transitive in the sense defined here and given to mean that a set of preferences 'remains the same over time'. Little defines 'consistency of choice' to mean transitivity 'together with invariance of choice between every pair of alternatives'. Hicks uses the term transitive in the normal way and consistency or 'an unchanged scale of preferences' is with him a feature of an ideal consumer 'who is not affected by anything else than current market conditions'. Apparently a deterministic view is taken so that a consumer cannot change preferences, and since market conditions leave preference orderings unaffected the ideal consumer's choices 'always express the same ordering'.[15] This irregularity of usage may have something to do with the use of axiomatic constructs in economics. Such constructs are timeless so that invariance over time is ruled out. Consistency and transitivity may therefore be used interchangeably. They are properties of functions and concern such mathematical questions as integrability. One cannot have any other interest in the matter. In a temporal context, on the other hand, transitivity and consistency in the sense of invariance over time cannot always be distinguished from each other. If a consumer is seen first to prefer A to B, then B to C and then C to A, we may say either that this preference ordering is intransitive or that the consumer has changed the preference ordering during the course of the observation. In either case, however, the consumer would behave irrationally in Becker's sense, and that is really the question I want to investigate.

When we assume that more profit is always preferred to less we assume a partial preference ordering which is both transitive and consistent and anyone who reveals such a preference ordering can be said to act rationally according to the consistency criterion. Similarly, where separate maximization of the quantities of goods is possible and always carried out, we may also speak of rational action in the consistency sense. When separate maximization is not possible then, as we have seen, the assumption that more is always preferred to less cannot explain choice. All we can say is that the most preferred collection of goods is chosen, though we have no rule for determining which is the most preferred collection. At the end of section 4.1 we saw that when such a rule, or the content of preferences, is removed we are left simply with the notion of purposeful action. The question now is whether purposeful action in this sense is equivalent to rational action in the consistency sense. It may be seen that it is not. Rational action in the consistency sense requires us to assume that there *is* a rule for determining what is most preferred even though we may not know the rule, whereas purposeful action requires no such assumption. A vestige of the content of preferences thus remains in the former but not in the latter. One may put this another way. Purposeful action requires us to assume merely that people have preferences whereas rational action in the consistency sense requires us to assume that people *always* have the *same* preferences.

It may be objected at this stage that I have put an incorrect interpretation on the observation by Becker quoted at the beginning of this section. Economists do not regard a change of tastes as a sign of irrational behaviour. I think that this objection may be only partially sustained. One must be careful here not to use axiomatic constructs for performing conjuring tricks. We have seen that when an economic model is an axiomatic construct, consistency or transitivity is a property of mathematical functions. A change of tastes, on the other hand, is a change in the axioms, or in the data, if one likes to put it that way. Whenever such a change is made one may assume that the new preference functions also are consistent or transitive. In a temporal context, however, the matter is more complicated. Even if one is dealing with hypothetical choices, they are then successive choices and, since transitivity involves a comparison of at least three choices, a transitive preference ordering must last at least for some time.[16] One may think of a consistent and transitive preference ordering as a structure in the sense that there are constant relations between its parts. The structure, representing tastes, may change occasionally, but only occasionally for if it changes very often transitivity loses its meaning. One may therefore amend the consistency criterion for rational action. Preference orderings do not have to remain the same always, but nevertheless for reasonable lengths of time.

Let us consider whether rational action in the consistency sense really conforms to what we normally understand by rational action. The proposition that at prevailing relative prices a majority of the English prefer tea to coffee, whisky to brandy and beer to wine, may or may not be consistent with what actually happens. One cannot, of course, observe preferences directly. We may infer some preferences though from the ratios between the quantities of goods sold. If relative prices do not change and the ratios between the respective sales turnovers of various goods in some area remain more or less the same over a lengthy period, we may conclude that the proposition that the people there have consistent preferences is not falsified. But what of the proposition that the English act rationally? We may or may not wish to contest this proposition, but we would not be inclined to judge it by the ratios of goods sold. One can surely act rationally without having habits or adhering to customs and fashions. Is consistency of choice then a necessary condition for rational action?

4.3 Means and ends

Another objection may now be raised. Am I not putting too much emphasis on the consistency of preferences? The 'consistent maximization of a well-ordered function' involves not only ends sought but also the means for attaining the ends. In this section I shall consider some ideas on the means ends relation while merely indicating some points that may be relevant in meeting this objection. I shall give my answer to the objection in section 4.4 in which I shall attempt to draw various threads together.

Let us begin by having a look at how Pareto dealt with the relation between means and ends. In the *Manuale* Pareto explained that economics is the study of 'logical actions'. Empirically, one seldom finds purely logical actions and the logical person is therefore an abstraction like the economic person. Though he gave very detailed illustrations of non-logical actions, he confined his remarks on logical actions to saying that there are objective relations between things of the form AB and subjective relations between mental concepts of the form A'B' and in logical actions A'B' conforms to AB.[17] In his later and more extensive *Trattato di Sociologia generale* he enlarged upon this. 'Suppose we apply the term *logical actions* to actions that logically conjoin means to ends not only from the standpoint of the subject performing them, but from the standpoint of other persons who have a more extensive knowledge'.

Logical and non-logical actions are therefore 'distinguished not so much by any difference in nature as in view of the greater or lesser fund of factual knowledge that we ourselves have'. Pareto acknowledged that the distinction was relative to the observer and remarked: 'One cannot imagine how things could be otherwise.'[18] One of the examples Pareto uses is that of ancient Greek sailors plying their oars and making sacrifices to Poseidon. While the sailors themselves may have regarded both actions as means for getting from A to B, the observer with the 'more extensive knowledge' would regard only the former as such. Actions may be purposeful without being logical in an objective sense.[19] To Pareto, then, logical action was action that would seem to be to the point to a good positivist. He focused attention on, so to say, the technological aspect of rational action.

It was this technological aspect of rational action that Schumpeter appeared to have in mind when he described the development of capitalism as the development of a rationalistic civilization.[20] Even von Mises, whose ideas on this, as we shall see, differed greatly from Pareto's, stressed this aspect. 'Man is in a position to act because he has the ability to discover causal relations. ... In a world without causality and regularity of phenomena ... man would be at a loss to find any orientation and guidance.' Von Mises explicitly opted out of the problem of what he called 'imperfect induction'. It was not important because an 'action unsuited to the end sought' is still rational though ineffectual.[21] However, it was this area that interested Pareto. It suited his purposes to be able to draw parallels between, for instance, what we call the instinctive behaviour of insects and the conduct of a human. Apart from a rational interpretation of action, he also put a behaviouristic interpretation on what humans do. More important, he put such a non-logical interpretation on conduct which, in his eyes, also had a veneer of rationalization over it. 'Human beings have a very conspicuous tendency to paint a varnish of logic over their conduct.' Hence his distinction between logical action, which the observer judges by what the observer believes to be the state of nature (which Pareto did not regard as absolute or final), and subjective purposes, for which the observer has to surmise what the actor regards as appropriate means to an end.

I shall return to the subject of the interpretation of action later. Here it is important to see that concentration on means does not exempt one from ends and the question of their consistency. First, Pareto's positivist observer would have to be able to distinguish between means and ends. If the ancient Greek mariners Pareto invoked had derived religious fulfilment or exhilaration from making sacrifices to Poseidon, or if they had simply found such ceremonies fun, then these sacrifices would not have been means to an end but an end in itself, and their actions would have been quite logical. Furthermore, there is the problem of consistency. Since Pareto spoke only of single ends or aims and not of sets of preferences, there can be no question of logical consistency (that is, transitivity), but the problem of consistency over time remains. Suppose the mariners set off for the Etruscan coast and, while still on their way, decide to stop for some fishing, and thereafter decide to turn around and head for the island of Rhodes instead of the Etruscan coast. A positivist observer would have to be able to keep up with all these changes of plans, otherwise the observer would get quite a wrong impression of the logicality of the actions.

It was a point such as this that was at issue in an article by C. Tagliacozzo which referred to a debate that had apparently taken place between Pareto and the philosopher Benedetto Croce.[22] Croce criticized Pareto's view that economics, which to Pareto, it will be remembered, is the study of logical actions, has much in common with mechanics. Croce's argument, as I interpret it, was something like the following: Pareto's logical action requires one to conceive ends as given facts. Croce illustrated the sort of cases in practical action in which one does conceive ends in this way by his Rhine wine example. If an individual has set itself an expenditure programme which does not include the purchase of wine and nevertheless succumbs to the 'temptation of the moment' and buys and drinks a bottle of Rhine wine, then this action will be followed by a judgement of self-disapproval (for being weak-willed) – something completely unknown in mechanics. Without such disapprobation or other appraisal, the notion of a *set* expenditure programme has no meaning in practical action; the purchase of Rhine wine would simply be the expenditure programme of the moment. Tagliacozzo then interprets economics as the study of the allocation of means to given ends or to a set expenditure programme. Such a programme does not have to be a long-term programme. 'Instead of referring to a year, a month, a week or a day, it may refer to an hour or an instant.' In an instant, however, means and end, plan and objective, are one and the same intentional act. What is done is both the end sought, the temptation of the moment, and also the means for attaining it. As the duration of preference orderings is shortened, Tagliacozzo argued, so economics approaches a limit in which there is, as Croce saw it, 'real action' as distinct from economic action.

Let us consider the views of another writer. Robbins's well-known definition of economics is in terms of means and ends, and is therefore relevant in the present context. (Subsequent page numbers refer to Robbins's 'Essay'.[23]) Robbins acknowledged the influence of Max Weber's conception of a sociology without value judgements (p 90). His insistence on the givenness of ends (pp 24, 46, and so on) may therefore merely have emphasized that economics does not itself make value judgements. Nevertheless, one may ask whether the givenness of ends in his scheme also held any implications about the duration of preference orderings. He did say that ends may be 'defined and understood' (p 24) and he even went so far as to announce a direct analogy between pure mechanics and pure economics (p 83). This is what Croce objected to in Pareto, and Robbins was tackled on similar grounds.[24] On the other hand, Robbins stressed that economists have no means for determining the movement of relative scales of valuation, that is, of preference orderings, and significantly he called this the 'irrational element' in economics (p 126). His remarks about elasticity of demand calculations implied that he did not think that such movements were infrequent. 'Is it possible reasonably to suppose that coefficients derived from the observation of a particular herring market at a particular time and place have any *permanent* significance – save as Economic History?' (p 108, his italics). In what sense then did he conceive consistent and rational action?

Consistency was first used in the sense of transitivity of preference orderings (p 92). Towards the end of the book, in the chapter on the significance of economics, consistency or rationality (used synonymously) assumed a different meaning, or rather two related meanings. First, it meant awareness of opportunity costs. 'For rationality in choice is nothing more and nothing less than choice with complete awareness of the alternatives rejected' (p 152). Second, it meant compatibility of simultaneous aims (pp 155f) According to Robbins, economics by no means assumes that people in general act rationally in this sense, but it 'provides a technique of rational action' (p 157). In other words, economics was here seen as a logic on which an individual may base its conduct. Now, if a person reasons in terms of alternatives foregone or in terms of maxims such as that bygones are bygones, then it does not matter whether others do likewise. It merely means the person reasons. Some of the examples Robbins used, however, showed that he had something much more elaborate in mind. People can choose rationally between economic systems as long as they understand 'the essential nature of the capitalistic mechanism' and the 'conditions and limitations' of alternative systems, that is, as long as they are aware of the opportunity costs, and 'economic analysis' can provide the requisite knowledge (p 155). Clearly, this is not merely a matter of logic and of individual action. The way that economics explains the market mechanism certainly involves assumptions about the actions of a large number of people. The person who wants to use economics as a technique of rational action may have to assess the extent to which others are using that technique. In this assessment the person would face the difficulties that, as we saw, Pareto's positivist observer would have to overcome in judging logical action.

4.4 A short synthesis

It is time to draw various threads together. For this purpose let us consider a very basic notion of action. Von Mises was wont to speak of action as follows: 'Acting man is eager to substitute a more satisfactory state of affairs for a less satisfactory. His mind imagines conditions which suit him better, and his action aims at bringing about this desired state.'[25] This notion is surely familiar to everyone and we can take it as a starting point. One may pick three elements out of it for closer examination: First, there is the desired state or aim. Second, there is the original state or the circumstances in which the actor finds himself. Third, there is the knowledge which the actor applies to transform the latter into the former. Since this knowledge may well be considered fallacious by others I shall refer to it simply as a belief. We therefore have aims, beliefs and circumstances.

Our beliefs and circumstances are likely to be such that a number of states seem to us attainable. Our aims must therefore be the most preferred of these attainable states and for this reason we are said to have scales of valuation or preference orderings, even though we may not be conscious of them. One widespread notion of rational action, we have seen, requires that these preference orderings be consistent, and in a temporal context this must mean that they remain unchanged for reasonable lengths of time. Robbins's notion of rational action (which no doubt conforms more closely to common usage) requires our beliefs to be such that with a liberal use of logic we may have an accurate idea of what is attainable, that is, that we may know what we are choosing and what we are rejecting and that our aims do not contain elements which cannot be brought about simultaneously. Pareto's notion of logical action requires our beliefs to be such that we can really transform our present circumstances into our aims. When our beliefs do not meet these standards, our aims still rank as subjective purposes. We may act purposefully even if our actions are not Pareto logical and Robbins rational.

In the foregoing I have deliberately used the first person *we* and *us* because the notion of action is meaningful to us irrespective of how we interpret the behaviour of others. The three elements I have picked out of the notion of action are found in our personal experience. We aim, we believe and we experience our circumstances, even when our aims are momentary whims and we engage in 'real action' in Croce's sense. The matter is quite different when we assume the role of observers and apply the notion of action to the behaviour of others. The circumstances in which an individual finds herself or himself are partly physical (attitudes to her or his environment and to other people are not) and in so far an observer can ascertain them. But the observer can have no direct experience of the individual's beliefs, aims and preference orderings (which also reflect attitudes to the environment). By questioning and surmise the observer may well build up a picture of the beliefs and preferences of the observed subject. The information on which the observer must base the surmise consists of certain aspects of the subject's original circumstances and the outcome of the subject's action. The observer must therefore reason backwards from the outcome of action to the original circumstances in order to surmise the observed subject's beliefs and preferences. The observer draws on personal experience of action for the surmise, but the views the observer takes of his or her own action and of that of others are nevertheless quite different. In the observer's own action the observer takes a *forward* view from her or his circumstances, beliefs and preferences to the outcome of action; as an observer she or he reasons *backwards* from the outcome of action to the subject's known circumstances, and in the process the observer surmises the subject's aims, preferences and beliefs.

There is an unlimited variety of beliefs and preferences that could possibly link the outcome of action with the actor's original circumstances. In order to narrow down the field, the observer has to make assumptions. The assumption that an action was Pareto logical, that is, that the outcome of an action was also the desired state or aim, delimits both the aim and the beliefs at the time of action. The further assumption that the action was Robbins rational, that is, that all the alternatives were calculated, also throws some light on the actor's apparent preference ordering at the time of action. Rational action is here an assumption. If the observer wanted to surmise whether an action really was rational, the observer would have to know the subject's beliefs and preferences and cannot simply surmise both whether an action was rational and also the beliefs and preferences that led to it. Since we are continually engaged in such surmises in our everyday social relations, we probably acquire considerable expertise at them. With the use of questioning, quite satisfactory interpretations of action are no doubt reached. But questioning also requires assumptions. The actor must be able to articulate his or her preference ordering and relevant beliefs, and must be able to remember them correctly. Also, the answer to a question is itself an action that requires interpretation.

I shall now try to apply this analysis to the kind of deterministic economic model I discussed in Chapter 3. It seems to me this model is based on an analogy with the forward-looking view of the personal experience of action or volition. All the conditions for the outcome of action or for the final state are already inherent in the initial state, that is, in the circumstances, beliefs and preference orderings of individuals. (For example, comparative statics seems to be based on the assumption that once an equilibrium has been disturbed all the conditions for the new equilibrium are already in existence and need only rational action for their realization.) On the other hand, the model does not refer to the action of an individual, but to the actions of a large number of people as seen by the economist observer. As an observer of action, the economist establishes a link between the initial state and the final state (the outcome of action) by assuming rational action. As we have seen, this is normal procedure for an observer of action, but in this case the economist cannot reason backwards from the outcome of action because it is this that the model is meant to determine. The economist is therefore forced to make assumptions also about the actors' beliefs and preference orderings. If action is Robbins rational and Pareto logical, the actors' beliefs cannot be idiosyncratic. Rather they must be such that, first, all the alternatives that exist in some real sense are known so that aims are really the most preferred of all possible aims, and, second, all aims are really achieved or all plans succeed. Robbins rationality and Pareto logicality therefore set objective standards for individual beliefs and one can therefore refer to them rather as a common fund of knowledge.

Now, the propositions of logic and of mathematics are agreed to by all, and one merely has to assume that they are really used. For empirical knowledge we may turn to the five data sets I listed in section 3.5.6 of Chapter 3. Technical constraints are usually expressed as the state of technical knowledge which, while not final, is common to all. Property norms, laws of contract and the prevalent forms of affiliation and coalition are also assumed to be known to all. However, the actors' knowledge of each other's initial circumstances and preference orderings creates difficulties peculiar to analysis in terms of deterministic economic models. The rule that more is preferred to less is taken as a commonly recognized guide to action, but, as we have seen, it cannot cover all cases.

If an individual does not know or is mistaken about the preferences of others his false trading may introduce fortuitous factors into the course of events that may destroy the deterministic premise that conditions for the final state are already inherent in the initial state, and need only rational action to be realized. For the model to be maintained, it must be assumed either that all individuals somehow know each other's circumstances and preferences so that each may choose that most preferred aim which is consistent with a common outcome of all actions or that plans are revised through a *tâtonnement* process according to preference orderings which already permeated the initial state as a kind of underlying network. In either case, preferences have to display some temporal consistency. If a preference ordering were a mere will-o'-the-wisp that changed with the temptations of each moment, it could not serve as a guide to action nor be 'discovered' in the market.

I want to suggest that the difficulties peculiar to analysis in terms of deterministic economic models stem from the attempt to combine the detached view of an observer with the personal experience of an aiming, forward-looking action. The constructor of such a model approaches the subject matter as an observer but, unlike other observers of action, does not want to infer aims and beliefs from the outcome of action, but ideally to infer the outcome of action (for example, an equilibrium price) from aims and beliefs. The economist can, however, have no direct experience of the aims and beliefs, let alone preferences, of others.[26]

It is here that the assumption of rational action, in the consistency as well as what I have called the Pareto and Robbins senses, comes in useful. With such an assumption it does not seem to matter how one may become aware of people's aims, preferences and beliefs, for it can then be shown that the conditions for the outcome of action or the final state are already inherent in the initial state, that is, that there is nothing between the initial and final states that can happen by pure chance and so violate the deterministic premise. Not only are preference orderings the same in the initial and final states, but Robbins rationality and Pareto logicality set objective standards for the direction action takes with any concatenation of preferences in the initial state when rational action in the consistency sense is combined with what I have called the Robbins and Pareto senses (which otherwise may simply serve as aids in the interpretation of action); the notion of rational action does indeed give economic models some of the characteristics of the models of classical mechanics.

This brings us back to the difficulty noted at the end of section 4.2, namely, that our intuitive understanding of rationality does not seem to require us to have consistent or constant preferences. The issue may be resolved after a fashion if models are formulated as axiomatic constructs in which consistency is merely a logical property. That way the troublesome notion of time is simply banished. However it is not a solution that has satisfied everyone. Von Mises, for instance, took a stand against it. 'A logical system must be consistent and free of contradictions because it implies the co-existence of all its parts and theorems. In acting, which is necessarily in the temporal order, there cannot be any question of such consistency.'[27] There is another solution, and that is not to confuse our roles as individuals who execute action with our roles as observers who use the notion of action to draw up intelligible accounts of the behaviour of others.

4.5 Action as a logical category

The distinction made at the end of section 4.4 will be fundamental in the remainder of this study. It will lead back to the distinction between *ex ante* and *ex post* facts, for the notion of a guide to action is an intrinsic part of our personal experience of action, while our interpretation of the actions of others is perhaps the most common form of what I called the genetic understanding. Before proceeding with this line of thought, it will be necessary for us to consider how the notion of action evolved into a logical category in the hands of certain prominent members of the Austrian school of economists. I indicated in Chapter 2 that members of this school may be associated with the view that the empirical study of action must necessarily be an interpretation of past actions, that is, an explanation of the past. Since the founder of the school engaged in a methodological dispute – the *Methodenstreit* – with the historical school of Schmoller, the reader may have found this an unlikely or at least a curious association. The fact that there is no real contradiction must be understood in the context of some basic Austrian tenets.

The *Methodenstreit* concerned the respective merits of what in Windelband's later terminology would have been nomothetic and ideographic procedures in economics, that is, of a procedure that sought generalizations or laws (nomos) and of one that sought descriptions of individual phenomena (drawn pictures).[28] The view had gained ground during the 19th century that a human, unlike an animal, does not have a nature but only a history, and that the social sciences should not seek explanations in terms of human nature (as had been widely done in the 18th century) but should rather look to the evolution of social institutions. This tendency of thought manifested itself in economics, particularly in Germany but also elsewhere, in an historical approach. Classical political economy came increasingly under attack and was ridiculed for its simplistic views of the human. Among those who reacted against this trend were Jevons and Menger.

Menger set out his views in his *Untersuchungen*[29] (page numbers refer to this work in the original German). The study of economics, according to Menger, fell naturally into three divisions. The historical, the practical (he gave the example of Finance), and the pure or exact as he called it (pp 11–30). The exact and the practical studies required a generalizing procedure and he charged the historical school with wanting to replace these studies with the historical study and its individualizing procedure. Menger himself conceded a useful role to all three divisions, but by implication put the exact study in a pivotal position. The historical study required the concepts of the exact study as a means to its own ends (p 18) and if prediction and control were at all possible in the social sphere, the practical studies likewise would have to turn to the exact study for its concepts (pp 5 and 25–30).

The nature of Menger's exact study, however, is somewhat peculiar by modern standards and also by the standards of the times in which he wrote. It appears to have been influenced by ideas that had come through from the schoolmen of medieval times. He spoke of regularities but did not mean by these empirically derived regularities. He conceded, rather patronizingly, that the (what he called) realistic-empirical direction of research may possibly establish some empirical generalizations in the social sphere but these would not be exact nor necessary regularities; they would not be laws of thought (p 40), a point that von Mises was later to take up. Both the exact and the empirical were theoretical as distinct from historical, but the exact study could get to the intrinsic nature of exchange, price, supply, demand and so on, by examining 'types' and the relation between types. Its conclusions are then universally true even if the types are never found in a pure form in experience (if they are, they are 'strong types' and empirical study may come to the same conclusions as the exact). In Appendix VI (pp 262–6) for instance, he argued that there is only one path between given means and ends that is logically the most efficient (*zweckmässigste*) and that this is the exact or economic path. The actual path that people follow may deviate more or less from the exact and the latter can therefore not be found by empirical means, or only vague approximations can be found. The idea therefore that the theorems of exact economics can be tested, as we would say nowadays, seemed to Menger as ludicrous as testing the theorems of Euclid by drawing triangles (p 54). In the end, though Menger disliked the implication of arbitrariness, exact seemed to mean little more than abstract. History tries to make us understand *all* aspects of certain phenomena, while exact theory tries to make us understand *certain* aspects of *all* phenomena (p 67). Moreover, exact theory later appeared as a kind of abstract form of history. Having considered what others have called the conjectural history of money (barter is a nuisance, and so on, pp 172ff), of markets and of other institutions, he came to the conclusion that the exact analysis of the origin of spontaneous, that is, undesigned, institutions is methodologically identical to the analysis of the determination of market prices, wages, rates of interest, and so on (pp 182–3). The latter are also the unintended consequences of purposeful action, just as a commodity money that has arisen spontaneously.

It was little wonder then that von Wieser began an exposition of the Austrian theory of value for English readers in 1891 by pointing out that the German historical schools and the abstract Austrian were far more closely related than it may at first have seemed.[30] Von Wieser took over Menger's epistemology more or less intact, as far as one can see, though he did not have quite the same breadth of view. It was von Mises who put Menger's ideas in a new light.[31] Von Mises adopted, or was influenced by, an epistemological outlook which in philosophical circles would be called neo-Kantian, and he sided with that branch of this movement which actively opposed positivist methods in the social sciences. With this background it was not surprising that he should have felt that Menger's ideas on types found in thought could be formulated much more precisely if we recognized our personal awareness of action as an *a priori* logical category on Kantian lines.

Von Mises was much criticized for his *apriorism*, but many of his critics misunderstood him. They had in mind an analytic *a priori*, that is, an arbitrarily chosen definition or axiom. Von Mises had in mind Kant's synthetic *a priori* which, according to that philosophy, is not arbitrary. Kant argued that the proposition 5+7=12 is not arbitrary. The sounds and symbols we use to express it may be arbitrary but not its meaning. If we write arbitrarily 5+7=13, then the sound and symbol for 13 would have the meaning of 12 (or five and seven would have other than their normal meanings) since we cannot conceive the matter in any other way. Nor can we discover the proposition 5+7=12 empirically. No one can combine five articles and seven articles and then *discover* that there are 12, unless one can already count to 12. Knowledge of numbers is in this sense prior to experience; it has to be taken to experience and is therefore an *a priori* category (as are space, time, causation, and so on). Von Mises conceived human action in the same way. It is not something we distil from data, but something we use in order to formulate facts in economics. I shall let von Mises speak for himself.

> If we qualify a concept or a proposition as a priori, we want to say: first, that the negation of what it asserts is unthinkable for the human mind and appears to it as nonsense; secondly, that this a priori concept or proposition is necessarily implied in our mental approach to all the problems concerned, i.e. in our thinking and acting concerning these problems.

We cannot 'interpret our concept of action as a precipitate of experience. It makes sense to speak of experience in cases in which also something different from what was experienced *in concreto* could, have possibly been expected before the experience'. Our cognition of action is simply 'the cognition of the fact that there is such a thing as consciously aiming at ends'. 'All the elements of the theoretical sciences of human action are already implied in the category of action and have to be made explicit by expounding its contents.'[32]

The science that explores the *a priori* category of action von Mises called praxeology and economics is apparently its only developed branch. Like Menger's exact science of economics, praxeology cannot benefit from empirical research any more than logic or mathematics can. If means are sufficient for A or for B but not for both, then the one not chosen is the other one's opportunity cost. It may not be a profound insight, but it cannot be confirmed or disproved empirically. One may show that a particular A and B are not really mutually exclusive, but this does not alter the logic or the meaning of opportunity cost. More generally, the category of action provides the concepts for the cognition of empirical economic facts and so the latter can never reveal new praxeological insights. Von Mises pointed out that all the familiar concepts of economics derive from the category of action. Scarcity, choice, preference, gain, loss, profit, cost, value, capital, success, failure, and so on, could have no meaning to a being who was not aware of volition or action, and cannot be applied sensibly to the analysis of any entity unless we interpret its behaviour as purposeful action. Concepts such as money, price, wage, market, and so on, require the (empirical) existence of certain institutional arrangements, but, as Menger already reasoned, these institutions also must be understood in the context of action. Hayek also has argued that cognition of such institutions depends on concepts provided by the theoreticians of the social sciences.[33]

Von Mises divided the science of human action into two branches; praxeology which is *a priori*, and history which is concerned with empirical facts derived from the interpretation of action (pp 41–6 of the *Foundation*, see note 32). He subscribed to the view (with at least one exception to be noted later) that human action is unpredictable – his remarks about attempts to find regularities in economic statistics are scathing – and thus history is the only empirical study of human action. The relegation of the empirical study of action to history was to von Mises, as it had been to Menger, quite compatible with a theory of action, for theory was not concerned with empirical facts, though it provided the necessary concepts for them, nevertheless, praxeology and history each had a distinct status in his eyes.

It is here that one begins to suspect that von Mises's neo-Kantian approach, if it is such, is of an unusual brand. It was precisely Kant's point that the *a priori* categories are barren in themselves and that one should study them for the insight they give into the nature of our empirical knowledge. Put plainly, the question is how do we use praxeology (which really means economics) if it is *a priori*. 'Into the chain of praxeological reasoning', said von Mises, 'the praxeologist introduces certain assumptions concerning the conditions of the environment in which an action takes place. ... The question whether or not the real conditions of the external world correspond to these assumptions is to be answered by experience' (p 44). Though he made many such remarks, he was never very explicit about the kind of conditions that could qualify for such assumptions. Since he regarded the facts of action as unique history, such an explanation seems not out of place. If the environment is a social one, can one avoid assumptions about consistent preferences – which von Mises did not believe in? Furthermore, anyone who can make experience answer questions about the realism of assumptions, that is, anyone who has mastered the art of induction, may well regard the praxeological reasoning which is to follow as child's play hardly worth mentioning.

Kirzner, who is very sympathetic towards von Mises's ideas, gives a down-to-earth example of praxeological reasoning. A city is served by alternate means of transportation. When one of them is put out of action, an observer will know that the other 'will tend to be employed in larger than normal volume'.[34] It is no doubt a conclusion one may come to without ever having heard of praxeology, but, as Kirzner points out, it is based on assumptions about human purposes. It may of course be quite wrong. People may stay at home or, if the services in question are a train and a bus service and the latter breaks down, it may well be that people take to their own cars since they expect the trains to be crowded and it will be the roads and not the trains that are congested. The point is that the reasoning surely presents no difficulties: it is knowing what assumptions to make that is difficult. Once we feel entitled to make certain generalizations about preferences and beliefs, the rest is easy. Robbins, whose means–ends definition of economics owes much to Menger-Mises ideas, also offers some examples of how assumptions are inserted into economic reasoning. If we want to show the effects on price of the imposition of a tax, we make suppositions about the elasticity of demand and the cost functions, and then the 'conclusions are inevitable and inescapable'. They must be. 'They are implied in the original suppositions.'[35] But if it is all in the assumptions already, then surely the crux of the matter is making assumptions and not reasoning from them. As we all know, newcomers to economics are inclined to say that they expected the subject to show them how to arrive at useful suppositions, and that they would gladly have supplied the logic themselves.

The criticism that logic is stressed at the expense of the cognition of fact is applicable not only to von Mises's scheme, as Robbins's example shows. However one wonders why von Mises, having introduced an *a priori* concept of action into economics, did not use it more effectively. All that has been said about *a priori* categories over the last 200 years has had little to do with inserting assumptions into logical schemes. *A priori* categories are interesting only when they are used to tell us something about the nature of the facts about which we make assumptions.

4.6 Beliefs and preferences in *ex post* and *ex ante* contexts

In section 4.4 I stressed the difference between the retrospective view we take up when we account for the occurrence of an event by interpreting the actions of others and the prospective view we take up in our own actions. In section 4.5 I argued that an *a priori* category of action is useful only when it tells us something about the nature of what we perceive as empirical facts. I shall now try to show the following: When we treat the beliefs, aims and preferences of others as empirical facts, the nature of such facts differs according to whether they are elements in our intelligible accounts of the actions of others or whether they are factors that we take into account in our own actions, that is, whether they are guides to our own actions. Conformably to the terminology developed in Chapter 2, I shall refer to beliefs, aims and preferences in *ex post* and *ex ante* contexts.

I shall deal with the *ex post* context first. In an early article von Mises stated the position well.[36] Action takes the form of a choice between alternatives. Only the visible effects of choice are given to the observer. He grasps the meaning of choice by positing the concept of relative importance or preference. When an individual chooses A rather than B, the observer takes this as an indication that, in the *moment* in which the choice was made, A appeared more important, more valuable or more desirable than B. Von Mises pointed out that one cannot here make a distinction between economic and non-economic conduct. The whole act must be interpreted, whatever the valuations behind it. One is concerned with a concrete, individual and possibly unique situation. We saw in section 4.4 that an interpretation of action must also include surmises about the acting individual's beliefs, that is, about the individual's expectations and what it regards as knowledge. Many variant interpretations may therefore be possible. Furthermore they are interpretations based on the belief that the observed individual did act purposefully. A behaviouristic interpretation or one in terms of Pareto's non-logical conduct may also be possible. It should be noted also that an assumption of consistent preferences is here not necessary. The surmised beliefs and preferences relate to the moment in which a choice is made. The observer cannot expect the facts at which he or she arrives to serve as a guide to action – they are parts of an account of how a particular event came about and there is no presumption that the acting individual has *necessarily* revealed any propensities that will shape the acting individual's future conduct. It can be seen from all this that beliefs, aims and preferences in this context are what I called *ex post* facts.

I turn now to the *ex ante* context. Our awareness of our own purposes involves aims and certain beliefs about how we can attain these aims. The beliefs, aims and preferences of other people are for us, as Pareto put it, among the obstacles in our way. When we feel justified in believing that any of these beliefs, aims and preferences will remain unchanged at least for some time they become guides to action or objects of our own beliefs. The old rule that people prefer more to less of a certain thing is no doubt widely applied when plans for action are conceived. Even von Mises, who regarded human action in general to be unpredictable, had to admit that there is 'no action that could be planned or executed without paying full attention to what the actor's fellow men will do'. The actor has to have some means of 'anticipating the conduct of his fellow men'. Almost apologetically, he said: 'Out of what we know about a man's past behaviour, we construct a scheme about what we call his character.'[37]

Apart from personal characteristics and habits, there are also preferences which extend over groups of people and one then speaks of fashions and customs. The word custom also refers to institutionalized means rather than ends and includes what in Chapter 3 I called the known forms of affiliation and coalition. The individual actor may base his or her plans of action on these in the belief that others are also doing so. The belief, for instance, that others are trying to make their purchases and sales in conventional markets is used by the individual actor as a guide in planning his or her own purchases and sales. I do not wish to pursue this subject in this place, though it deserves a closer examination than it has usually received in economics. The point I want to make here is that beliefs and preferences (of others) in an *ex ante* context are quite different from those in an *ex post* context. They are used in much the same way as are facts established in the natural sciences. Max Weber repeatedly stressed this. An individual has to assess the likelihood of achieving various aims, with expectations based on regularities, maxims and rules of thumb. It does not matter whether the latter relate only to physical processes or also to the actions of other people.[38] Weber typified a maxim in terms very similar to the definition of a causal fact in this study. He pointed out that the actions of a person taking on employment and accepting payment in cash must be based on such maxims, that is, on expectations of the actions of the employer, shopkeepers, and so on. An entrepreneur bases production plans as much on the expectations of the actions of employees, customers, magistrates, and so on as on purely technical knowledge.

No doubt such maxims are often elaborated into something akin to the models of economic theory. For instance, the belief in the rule that more is preferred to less, together with some other considerations, may lead to the expectation that a reduction in the size of the labour force in a certain area will bring about higher wage rates. Nevertheless, these maxims are isolated *ex ante* facts; their context is the personal awareness of a need for knowledge that may be applied in the conscious aiming at ends. The use of combinations of these maxims as deterministic models, in which the economist is a purely passive observer of the actions of others, may not be warranted.

Notes

1   W.S. Jevons, *The Theory of Political Economy*, 5th edn (New York: Kelley & Hillman, 1957; first published 1871) p 16. He mentioned two more 'simple inductions' 'known to us immediately by intuition', namely, 'that human wants are more or less quickly satiated: that prolonged labour becomes more and more painful' (p 18).

2   J.S. Mill, *A System of Logic*, 9th edn, vol II (London: Longmans, 1872; first published 1843) p 494. (See note 23, Chapter 2.)

3   A. Marshall (*Principles of Economics* 8th edn, [London: Macmillan, 1920] p 14) retained much of the older view. Economics 'concerns itself chiefly with those motives which affect, most powerfully and most steadily, man's conduct in the business part of his life'. Of course, people also bring 'high ideals' into business. 'But, for all that, the steadiest motive to ordinary business work is the desire for the pay which is the material reward of work.' However, Marshall added a neo-classical twist. Virtually any motive, even the most altruistic (p 16), could be subjected to 'money measurement'. When Pigou (*The Economics of Welfare*, 4th edn [London: Macmillan, 1960], p 11) said that he would restrict himself to the social welfare 'that can be brought directly or indirectly into relation with the measuring-rod of money', he probably also thought in terms of monetary valuations of motives. Mill, in common with most of the better-known of the classical political economists (Senior was a possible exception), did not, I think, see the matter in this way. The desire for wealth or for monetary gain was just one of many known motives, and the others did not concern political economy.

4   This statement may suggest that there was a greater unanimity of outlook than there actually was. Many political economists saw their subject as a study of material wealth, which could be conducted without any reference to choice or preferences. It was this definition of the subject that Lionel Robbins attacked in his famous 'Essay'. (L. Robbins, *An Essay on the Nature and Significance of Economic Science*, 2nd edn [London: Macmillan, 1935], pp 4ff.) However, the view that the subject matter of economics consisted of an aspect of human conduct rather than of material objects increasingly gained ground during the 19th century. It came to fruition in 'subjective' theories of demand, but was already quite marked in the latter days of the classical period. Certainly, J.S. Mill had already adopted an 'aspect' definition. The changing ideas on the demarcation of the economic subject matter have been explored at great length by I.M. Kirzner, *The Economic Point of View* (Princeton: Van Nostrand, 1960).

5   Mill, *op. cit.* (note 2) p 497. Mill did, however, acknowledge that political economy had to take into account two other motives. Political economy 'makes entire abstraction of every other human passion or motive: except those which may be regarded as perpetually antagonizing principles to the desire of wealth, namely, aversion to labour, and desire of the present enjoyment of costly indulgences. These it takes, to a certain extent, into its calculations, because these do not merely, like our other desires occasionally conflict with the pursuit of wealth, but accompany it always as a drag or impediment' (p 496). We have already seen that consumption and the disutility of work create stock and flow problems in deterministic models; Mill's reference to a drag possibly made the same point.

6   *Ibid.*, p 499. See also Robbins, *op. cit.* (note 4) pp 80ff.

7   Jevons, *op. cit.* (note 1) p 21.

8   J.A. Schumpeter, *History of Economic Analysis* (London: Allen & Unwin, 1954), p 302.

9   See, for instance, J.K. Galbraith, *The New Industrial State* (London: Hamilton, 1967) pp 128–97.

10   Robbins, *op. cit.* (note 4) pp 84–6 and 95–6, discusses hedonism in economics: and concludes that it has played an insignificant role in practice, even though some economists made hedonistic claims. Edgeworth, for instance, urged the conception of 'man as a pleasure machine'.

11   Schumpeter, *op. cit.* (note 8) p 130.

12   J.R. Hicks, *A Revision of Demand Theory* (Oxford: Clarendon, 1956), p 42.

13   Robbins, *op. cit.* (note 4) pp 110–11, considers the possibility of people being indifferent to everything. He thinks it 'overwhelmingly unlikely', but if such were the case 'conduct' would be 'indeterminate'. In what sense, however, would it be conduct? People may then potter around aimlessly, but they would not act purposefully.

[14] Gary S. Becker, 'Irrational Behaviour and Economic Theory', *The Journal of Political Economy*, 70, 1962, p. 1. I am indebted to Peter Lewin for pointing out this article to me. See also P.A. Samuelson, *Collected Scientific Papers of Paul A Samuelson*, vol 1, edited by J.E. Stiglitz (Cambridge, MA: MIT Press, 1966) p 130. '[G]iven the modern economist's reduction of the rational behavior concept down to little more than consistent preference behavior.'

[15] J.E. Meade, *The Stationary Economy* (London: Allen & Unwin, 1965), p 28; I.M.D. Little, *A Critique of Welfare Economics*, 2nd edn (Oxford: Clarendon, 1957) p 24; Hicks, *op. cit.* (note 12) pp 18, 23 and 47.

[16] This, of course, has nothing to do with the 'order of consumption' and its relation to integrability which Pareto discussed in the Mathematical Appendix of his *Manuale* and which subsequently elicited further discussion.

[17] Vilfredo Pareto, *Manual of Political Economy*, first published 1906, translated from the French edn of 1927 by A. Schwier (London: Macmillan, 1971) pp 29–32 and 103. Non-logical actions are illustrated on pp 32–101.

[18] Vilfredo Pareto, *The Mind and Society*, translated from the 1923 edition by A. Bongiorno and A. Livingston (London: Cape, 1935) vol I, pp 76–7. A discussion of Pareto's definition of logical action may be found in Talcott Parsons, *The Structure of Social Action* (Glencoe: McGraw-Hill, 1937) pp 185–91.

[19] Max Weber made a similar distinction between the purposeful and the valid or logical in the sense of Pareto, namely, 'subjektive Zweckrationalität' and 'objektive Richtigkeitsrationalität'. See Max Weber, *Gesammelte Aufsätze zur Wissenschaftslehre* (Tübingen: Mohr, 1951) pp 432–8. Pareto's term logical action would appear to be a misnomer since judgements of fact are involved. Pareto did not insist on the term. In a footnote he expressed regret at having to use words to describe the idea; he would rather have used a symbol. He did insist though that non-logical did not mean illogical.

[20] J.A. Schumpeter, *Capitalism, Socialism and Democracy* (London: Allen & Unwin, 1943) pp 121–5ff, *passim*.

[21] Ludwig von Mises, *Human Action* (London: Hodge, 1949) pp 20–3.

[22] G. Tagliacozzo, 'Croce and the Nature of Economic Science', *Quarterly Journal of Economics*, 59, 1945, pp 307–29. The article is discussed in Kirzner, *op. cit.* (note 4) pp 169–72.

[23] Robbins, *op. cit.* (note 4).

[24] See Talcott Parsons, 'Some reflections on the Nature and Significance of Economics', *Quarterly Journal of Economics*, 48, 1934, pp 513f.

[25] von Mises, *op. cit.* (note 21) p 13. One could cite many other passages from von Mises's writings in which he expresses himself in very much the same way.

[26] It appears to me that Hayek made a similar point in his address 'Economics and Knowledge' (in *Individualism and Economic Order* [London: Routledge, 1949] p 39). 'Datum means, of course, something given, but the question which is left open and which in the social sciences is capable of two different answers, is to whom facts are supposed to be given. ... There seems to be no possible doubt that these two concepts of "data", on the one hand, in the sense of objective real facts, as the observing economist is supposed to know them, and, on the other, in the subjective sense as things known to the persons whose behaviour we try to explain, are fundamentally different and ought to be carefully distinguished.'

[27] von Mises, *op. cit.* (note 21) p 103.

[28] Menger distinguished between two directions of research; the one sought the cognition of individual phenomena, the other the general aspects of phenomena. The distinction was almost identical to one made some years later by Rickert. 'Empirical reality becomes nature when we view it with respect to its universal characteristic; it becomes history when we view it as particular and individual.' Rickert then contrasted the 'generalizing procedure of the natural sciences' with the 'individualizing procedure of history', and this

became his formal principle for the classification of sciences. See Heinrich Rickert, *Science and History*, first published 1902, translated from the 7th German edn by G. Reisman (Princeton: Van Nostrand, 1962) particularly pp 56–7. Unlike Menger, Rickert, who had a wide spectrum of 'cultural' sciences in mind, championed the individualizing procedures. Rickert's views became very influential in the early years of this century. Among those influenced, as Hayek pointed out in the preface to the English edition of the book cited above, were Max Weber and Ludwig von Mises.

[29] Carl Menger, *Untersuchungen über die Methode der Socialwissenschaften* (Leipzig, 1883, reprinted by the London School of Economics as vol II of *The Collected Works of Carl Menger*, 1933). The book has been edited by Louis Schneider and translated into English by F.J. Nock as *Problems of Economics and Sociology* (Champaign: University of Illinois Press, 1963). My references are to the original version.

[30] F. von Wieser, *Gesammelte Abhandlungen*, edited by F. von Hayek (Tübingen: Mohr, 1929) p 35.

[31] See L.M. Lachmann, 'From Mises to Shackle', *Journal of Economic Literature*, 14, 1976, p 56. However, Menger's influence on von Mises seems to me to be underrated in this paper.

[32] L. von Mises, *The Ultimate Foundation of Economic Science* (Princeton: Van Nostrand, 1962) pp 6, 8 and 18.

[33] von Hayek, 'The Facts of the Social Sciences' in *op. cit.* (note 26) pp 57–76.

[34] Kirzner, *op. cit.* (note 4) p 173.

[35] Robbins, *op. cit.* (note 4) p 122.

[36] L. von Mises, 'Vom Weg der Subjektivistischen Wertlehre' in *Probleme der Wertlehre*, edited by L. von Mises and A. Spiethoff (Leipzig: Duncker, 1931) pp 77–8.

[37] von Mises, *op. cit.* (note 32) pp 46–51.

[38] Weber *op. cit.* (note 19) pp 325–8, 350, 441.

Variant Conceptions of Preferences

5.1 New and old paradigms

It is always interesting to consider the works of those who pioneered a new conception. In a study which scrutinizes the presuppositions on which paradigms are built, such works are especially important because their authors naturally felt obliged to justify their ideas and their criticisms of older ideas by an appeal to what they considered common sense. Once a paradigm has been set there is less need for this because there is then a circle of people to whom the esoteric idiom is meaningful. An economist can speak of a price change accompanied by a compensating variation in income which leaves a consumer on the same indifference level, and the economist will be understood by colleagues, even though to a layperson with a cursory contact with economics it may seem quite silly. The academic innovator has to address readers who are expected to be as sceptical as such a layperson. For this reason I have paid a good deal of attention to such pioneering authors in this study, and I shall do so again in this chapter in which I shall consider variant conceptions of needs, wants, tastes or preferences. Throughout the chapter I shall consider whether these variants should be seen in an *ex post* or an *ex ante* context.

5.2 Marshall

Marshall contemplated the inclusion of demand and of subjective elements into economic theory with much less enthusiasm than Jevons and Menger had done. He rejected outright the contention that the 'Theory of Consumption' is the scientific basis of the subject (p 90; page numbers refer to the *Principles* – see note 17, Chapter 2). One may speculate whether this merely reflected his eagerness to stress the continuity of his economics with the older classical political economy, or whether he really believed, as he seemed to insinuate, that his contemporaries were rushing in where the political economists had feared to tread. He quoted with approval a statement by McCulloch to the effect that the satisfaction of a want or a desire is merely a step in a novel and a creative human pursuit (p 90). '[S]o far as the expenditure of private individuals is concerned', Marshall commented '[t]he common sense of a person who has had a large experience of life will give him more guidance in such a matter than he can gain from subtle economic analysis'. The classical political economist had said so little about demand 'because they really had not much to say that was not the common property of all sensible people' (p 84).

However, he listed various reasons for the greater prominence of demand in the economics of his time and he then proceeded to deal with it. One could say, in the terminology developed at the end of Chapter 4, that Marshall saw demand in an *ex ante* context. I shall argue that it was to him one expression of what he called 'normal action'. The word 'normal' had a central place in Marshall's thinking. He explained that it was used instead of the word 'legal' when the law in question was 'a statement of relation between cause and effect' (p 34). He invoked the complexity argument to show that economic laws were estimates of 'tendencies of human action' (p 32) or 'statements of tendency'. Thus 'the course of action which may be expected *under certain conditions*' is normal action (p 34, his italics), that is, there is a tendency to such action. The example he used was that it was normal for a bricklayer to accept work at 10d an hour in most parts of England, but this was relative to certain conditions. 'In Johannesburg it may be normal that a bricklayer should refuse work at much less than £1 a day' (p 34). The conditions and therefore the norm had to be gathered from the context, and in this one was merely following the manner of the 'common discourse of life' in which the word normal takes on different meanings when, for instance, periods of different lengths are under consideration (p 363). Moreover, there was no sharp line of division between normal and abnormal conduct, just as there was not between normal prices and current market prices (preface to first edition). In market prices, 'the accidents of the moment exert a preponderating influence' but what are normal prices in relation to the current prices which change from hour to hour on a produce exchange are merely current prices in relation to 'the year's history'. Again the normal prices in the context of the year's history are current prices in the context of 'the history of the century'. Marshall's concept of normal is thus one that is meaningful to us all when, as individual actors, we cast about for guides to our own action. Few would have difficulties with an example Marshall uses to illustrate the use of normal. 'Illness is an abnormal condition of man: but a long life passed without any illness is abnormal' (p 34).

When Marshall considered changes in 'normal demand' (p 462) he said that, among other things, a 'great and lasting change in fashion', a new invention or the development of a rival product would make it necessary to make out a new demand schedule. The context suggests that one would not expect to have to make out a new demand schedule every minute or every hour, and that one did not have to rely for this on an assumption of consistent preferences convenient for drawing up a theory. Clearly, the demand curve reflected normal demand. But why did Marshall think that one could make a distinction between normal and the accidents of the moment in the case of demand? The answer, it seems, is that he was again casting his eyes back to his predecessors. Ricardo and his followers (according to Marshall) had emphasized, even overemphasized, 'that while wants are the rulers of life among the lower animals, it is to changes in the forms of efforts and activities' that we must turn in the analysis of man (p 85).

Marshall did not say precisely what he meant by 'activities', even though he devoted a short chapter to them and mentioned them frequently thereafter especially in the closing chapters of the book. They are best understood in the context of the Victorian belief in progress, that is, the belief that man constantly improves his condition in some sense or, as Marshall put it, that he develops forever higher activities.[1] The uncivilized man merely has biological needs, but with 'every step in his progress upwards' (p 86) he develops more activities and a greater variety of wants. For instance, 'in dress conventional wants overshadow those which are natural'. The demand for houseroom is influenced not so much by the need for shelter as by the desire for distinction, the development of 'social activities' and the need for privacy to cultivate 'higher activities' (pp 87–8). New wants arise in the pursuit of science, literature and art and in such pastimes as 'athletic games and travelling'. Those who somehow lagged behind in the development of activities evinced a singularly monotonous demand. In some parts of the world freed slaves (according to Marshall) spent their new freedom and wealth 'in idle stagnation that is not rest' and (again according to Marshall) a 'rapidly lessening part of the English working classes' spent on drink anything that was left after the bare necessities had been provided for. He concluded that while wants gave rise to activities in an uncivilized state, 'yet afterwards each new step upwards is to be regarded as the development of new activities giving rise to new wants, rather than of new wants giving rise to new activities' (pp 89–90). Normal demand or the current state of wants, therefore, appears to refer to a stage reached in the linear evolution of activities.

Having dealt with all this, Marshall proceeded with the demand analysis which is associated with his name. He stressed, however, that 'such a discussion of demand as is possible at this stage of our work, must be confined to an elementary analysis of an almost purely formal kind' (p 90). He noted that an implicit condition of the law of diminishing marginal utility is that insufficient time be allowed for an alteration of the character or taste of an individual (p 94). He did not tarry with the individual for long. There were no individual demand curves for wedding-cakes or for the services of surgeons, and in any case the economist was not interested in 'the variety and fickleness of individual action'. In large markets 'the peculiarities in the wants of individuals will compensate one another' and the market demand curve should then be taken to refer to a particular period – Marshall suggested a year (pp 93–9). The law of demand nevertheless was explained in analogy to individual demand curves, but the elasticity of demand was treated from the point of view of price statistics. At the end of his discussion of demand elasticity he appended a note on the statistics of consumption (pp 113f). Official statistics were not very useful for deriving demand curves because of *ceteris paribus* problems, but he recommended a suggestion made by Jevons that much could be gleaned from shopkeepers' books. A shopkeeper is likely to know of changes in his customers' incomes and in other conditions of their lives and could therefore judge the effects of price changes. For instance, if two winters are equally severe and rates of pay about the same but the price of butter different in each, Marshall thought that a shopkeeper would get a good idea of two points on a demand curve. The effects of price changes larger than those actually found could be estimated, he thought, by considering different income groups. A poor man's purchases at a price high by his standards could be used to gauge a rich man's purchases at a price which is high even by his standards. In piecing together information about what is presumably a single normal demand from such isolated observations and dubious comparisons between income groups, Marshall seemed to have gone well beyond the guidance which the common sense of a person with 'a large experience of life' is likely to provide.

5.3 Menger

Menger began his *Grundsätze* with a genuflexion to causation.[2] All things are subject to the law of cause and effect. It soon appears, however, that he separated a subjective sphere of human cognition and knowledge from the objective world and that causation applied only to the latter. One aspect of the objective world was that men had needs (*Bedürfnisse*) and that there were things which directly or indirectly (through further processing) had the capacity to satisfy such needs, or rather to bring about states that men regard as the satisfaction of their needs (pp 1–2). Men could have knowledge of the causal connection between their needs and the things that could satisfy needs, just as they could have knowledge of any other causal connection. When they did have such knowledge and had access (*Verfügung*) to the things that could satisfy needs, they regarded these things as goods (p 3). Needs and goods were, therefore, objective facts which one could get to know but which did not depend on arbitrary whim (*Willkür*, pp 85 and 121). However, as with any other knowledge, one could be uncertain or mistaken about these facts. When providing for the future, an individual was uncertain whether there would be a need, for instance, for medicines (p 36).

On the other hand, he might believe that he had a need which he did not really have, as when he imagined that he had an illness. Again, he might ascribe properties to goods, for example, medicinal properties, which they did not really have. In that case, Menger said, one had to speak of imagined goods (p 4). The more advanced a people were, the smaller the proportion of imagined goods to true goods. The distinction between imagined and true goods was of course similar to Pareto's distinction between logical and non-logical action. Menger pointed out that even Aristotle had made such a distinction.

So far the scheme of needs and goods is of little economic interest because there is no scarcity. Menger of course wanted to consider the economic cases where goods were insufficient to cover all needs (pp 51f and 77ff). Men then tried to provide for the future by gaining control over economic goods. For this purpose they had to be able not only to predict needs and to understand the properties of goods but also to judge the relative significance (*Bedeutung*) of the satisfaction of various needs and, therefore, of goods. They could then make up their provisions in such a way that the marginal significance of each good was equal (pp 87ff) – it is this aspect of Menger's work which is now regarded as his chief contribution. In judging significance, Menger argued repeatedly, men would first have regard for the maintenance of their lives and of those of their dependants. Beyond that they would take a longer-term view of their health and only then consider their feelings of well-being and enjoyment. The needs argument became more tenuous the further he moved away from biological needs. He did not argue in terms of activities, as Marshall was to do, but (which in some respects may amount to the same) stuck to the idea that well-being and enjoyment were causal effects of goods that had to be discovered and about which one might be mistaken. The significance of a good was an objective fact.

Von Mises later expressly repudiated the objective nature of needs.[3] The observer was given only the visible effects of action and it was futile to distinguish between what really were the acting individual's needs and what the individual thought they were. There is, however, an important difference. If needs and the significance of their satisfaction are facts which have to be discovered, maximization and the equi-marginal principle require conscious effort and enquiry and an observer may conclude that an individual's attempt to maximize has failed. If subjective wants are to be inferred from action, then whatever an individual does is compatible with, or must be interpreted as, successful maximizing actions, that is, the equi-marginal principle becomes a hermeneutic device and, if it is applied strictly, misapprehension on the part of the acting individual is meaningless.

Though in one place (p 85) Menger said that needs possibly were partly dependent on human will and habit, he usually saw them as facts which lay beyond the whim of man. For instance, making a Benthamite distinction, he said that men commonly erred in taking account only of the intensity of satisfaction without regard to its duration (p 122). The idea of the mind of a person contemplating in detached fashion the conditions of his physical body, may seem almost schizophrenic, but the distinction between the subjective and the objective manifested in it is an age-old presupposition. As Menger saw it, economics lay in the objective sphere, independent of the will of man (pp xlvif). Value, though, was a subjective imputation. As we have seen (section 4.5 of Chapter 4) he had not changed his mind 12 years later when he published his *Untersuchungen*. In Appendix VI of that work he argued that since needs and goods were part of the objective world, the 'exact' science of economics, which studied the most efficient application of goods to needs, is also objective. However, the subjective side also entered the picture because men erred and did not necessarily follow the most efficient course available. It was the subjective side which was of course to become the hallmark of Austrian analysis. Value, cost and so on cannot be understood without reference to the thoughts of economizing individuals and in this sense Menger's economics was also subjective however he may have described it. But to Menger the subjective side was made up only of human '*Erkenntniss*', that is, cognition and knowledge and not of human needs. Needs were later incorporated into the subjective sphere as wants. The change-over was not difficult. Needs about which men can be mistaken are for many purposes equivalent to wants about which by their nature (as something inferred from action or as something purely in the actor's consciousness) men cannot be mistaken.

Did Menger then take an *ex ante* or an *ex post* view of preferences? I shall argue that in effect it was an *ex post* view with certain *ex ante* aspects. The biological needs of human beings are not regarded as very changeable and in this sense there are consistent needs. Moreover, while satisfactions are experienced separately, humans make provision for their needs for some time ahead so that they are able to weigh up the relative significances of goods and equate them at the margin. Oskar Morgenstern was later to take some pride in this aspect of Austrian theory.[4] These are the *ex ante* aspects. But biological needs as treated by Menger do not make a deterministic model possible, nor can they even serve as reliable guides to action, for their consistency is vitiated by the subjective element, the fact that people make mistakes. In contrast with Marshall, Menger insisted that economics did not concern itself with practical suggestions for action to be taken (p xlvii). Instead, as he was to explain more fully in the *Untersuchungen*, it investigated the simplest elements or types needed to understand economic phenomena. Once they had been established they were to be used to explain the evolution of complicated economic phenomena (p xlv). Menger dedicated his book to Roscher and, at the end of his preface, he expressed his joy at being able to contribute to the then recent German developments in economics, which were thoroughly historical. His book, he said, should be regarded as a greeting from Austria. He wrote this before the *Methodenstreit* and I think it bears out what I tried to explain earlier. Even though the friendly Austrian overtures were rebuffed by the historical school, Austrian analysis was never far removed from a genetic explanation of economic events, that is, from a narrative account of the past.

5.4 Walras

Marshall wanted to see the question of preferences in an *ex ante* context. Menger, even though to him human needs were largely independent of a capricious will, nevertheless developed his scheme on the assumption that the question of choice would be considered in an *ex post* context. The practical Englishman wanted his subject to be useful for making business decisions and possibly for drawing up government policy. The Austrian, less interested in the hustle and bustle and problems of everyday decision-making, was content merely to explain the past and to follow an urge for understanding and order. Walras, it seems, could never quite make up his mind which of these two attitudes he should adopt, even though he appears to have been more conscious of the difference between them than were the other two. When he felt obliged to make a decision on this he usually opted for the *ex post* context.

With an endearing openness, Walras indicated to his readers that he was not too sure about the basic nature of human will. (Page numbers refer to his *Elements of Pure Economics*. See note 5, chapter 3). He announced, rather grandly: 'Alongside the many blind and ineluctable forces of the universe there exists a force which is self-conscious and independent, namely, the will of man.' Then, as though he thought he had expressed himself too strongly, he added: 'It may be that this force is not quite as self-conscious and independent as it supposes itself to be.' So he rephrased his original pronouncement. 'The essential point is that, at least within certain limits, the human will is self-conscious and independent' (p 61). With the question of free will considered but left inconclusive, how did he treat tastes or preferences? He took over from his father a distinction between extensive and intensive utility and gave the former a more definite meaning (pp 115–16). Extensive utility 'consists in' the amount demanded at zero price. William Jaffé pointed out (p 505) that Walras used the term utility in the sense of 'individual psychological reactions to goods'. Extensive utility, in common with Menger's '*Bedürfnisse*' or needs, was therefore given by the amount of a good that a person would require or want if it was not scarce.

As was the case with Menger's needs, extensive utility by itself did not get one very far. It established only one point on the demand curve. For the slope of the demand curve Walras brought intensive utility into play.

However, while extensive utility could be discovered, if one set price at zero, intensive utility was more troublesome. Walras considered the problem for a moment (p 117). At first glance it seemed impossible to pursue the analysis further because intensive utility was 'so elusive' and had 'no direct or measurable relationship to space or time, as do extensive utility and the quantity of a commodity possessed'. He then came to a decision which, if one considers the subsequent history of economics, was quite momentous. 'Still, this difficulty is not insurmountable. We need only assume that such a direct and measurable relationship does exist, and we shall find ourselves in a position to give an exact, mathematical account of the respective influences on prices of extensive utility, intensive utility and the initial stock possessed.' He therefore assumed 'a standard measure of intensity of wants ... applicable not only to similar units of the same kind of wealth but also to different units of various kinds of wealth' (p 117). This measure was to be 'rareté', his word for marginal utility. From the way Walras phrased this and from what he said subsequently, it seems to me that Walras had in mind an *ex post* context, an explanation of the past in which, if all the rest was known, intensive utility would appear as a residual influence. It was rather like positing an algebraic unknown and solving for it by inserting the values of all the other variables.

Once formulated, the concepts of intensive utility and rareté also had an appeal in an *ex ante* sense. He even toyed with the idea of making utility functionally related to time and thus arriving at 'economic dynamics' (p 117), which seems to suggest a deeper underlying structure according to which people behave. Nothing came of the idea. As is well-known, Walras went no further in this direction than a simple comparative statics. He considered this briefly in lesson 13, but had already broached the subject in lesson 10. There he said that the theorist had a right to assume that the determinants of price remained invariant over the period in which the price was established, but that he had to remember that they were by nature variable and that he had to 'formulate the law of the variation of equilibrium prices' (p 146). This, however, had nothing to do with making utility a function of time. It was merely a statement that a change in price must be due to a change in utility or quantity possessed (production had not yet been considered). These changes could be determined. He first suggested a direct investigation in which questions are put 'to each and every individual about the elements that enter into the making of his individual demand curve' (p 147). But then, as though this did not seem a very promising approach, he suggested that if a rise in price coincided with the discovery of a remarkable new property in the commodity concerned or with a catastrophe destroying part of the supply, one could not help associating these events with the rise in price. Since such a discovery or catastrophe would be a stochastic or chance event which is observed to coincide with a rise in price, he seemed definitely to be thinking in terms of an *ex post* explanation, a genetic understanding of a price change.

In lesson 22 Walras put the following remark into the mouth of an imaginary critic (p 256): 'One of the elements of the determination of prices under free competition is free will which entails decisions that are unpredictable.' He immediately countered: 'Actually, we have never attempted to predict decisions made under conditions of perfect freedom; we have only tried to express the effects of such decisions in terms of mathematics.' Clearly, he was not thinking of an exercise in comparative statics, that is, of the effects of a specified change. He was talking of the effects of *any* decisions, whatever they were. The effects, it turned out, were that within certain limits a 'maximum of utility' is attained if competition is free. For this conclusion, as can be seen, Walras did not require consistent preferences in the temporal sense nor even a normal demand in Marshall's sense. He did not require preferences in what I have called an *ex ante* context. Nor was he attempting an *ex post* explanation of a specific past event. The discussion of the logical form of this type of argument I shall leave over to the next and final chapter.

5.5 Choice without utility

In the later general equilibrium analysis for which Walras provided the inspiration, preferences were increasingly seen in an *ex ante* context. They were not put into the form of a normal demand as required by Marshall's partial equilibrium analysis, but rather into the form of the consistency or invariance of each individual's preferences. This development, it seems to me, had something to do with the attempted divorce of choice from utility or satisfaction. To Walras, as we have just seen, free choice still signified the attainment of satisfaction, and a maximum of utility had a normative implication. It makes sense to consider a possibly unique choice, made on the basis of a preference ordering which may last for only a moment, if it is taken to reveal what most satisfies the individual at that moment within the limits of his resources. When the link between choice and satisfaction, with its normative implications, was considered unnecessary, as Pareto considered it, some other significance must have been ascribed to the phenomena of choice. It was then taken to reveal a preference ordering which was somehow programmed into the mind of the individual, if not permanently then at least for some time. Understood in this way, choice provided information which could be used without any reference to utility or satisfaction, or indeed (once the information had been gathered) to any thoughts of the acting individual, as insisted upon by the Austrian school. The knowledge one gained by inspecting people's choices was no different in principle from the knowledge that a stone held above the ground will, when released, fall to the ground with a certain acceleration. There is no need to interpret the movement of the stone as an indication that the stone wants to be on the ground or that it is well satisfied when it gets there. The view of choice and action that was being adopted seems so exquisitely simple that it is a wonder that no one had thought of it before.

Although the term 'revealed preference' came much later, Pareto had already taken a decisive step towards the conception of choice and action outlined previously, especially in the Mathematical Appendix to the *Manuale*. In the text of the *Manuale*, choice was for the most part still tied to ophelimity, which was utility stripped of its normative connotations, and to indices of ophelimity. He also stressed constantly that economics was confined to *repeated* actions performed for the satisfaction of tastes. Since the subject was already confined to logical actions, its scope would therefore appear to have been much narrower than Pareto's discussion otherwise implied. When he came to consider for how long actions are repeated, that is, the duration of preferences, he used the example that Italians drink coffee rather than tea. Clearly, he was thinking not only of actions that are repeated by people individually but rather of customs – his mind was running along Marshallian lines – and he concluded that one could assume 'without perceptible error' that an indifference map would remain valid for up to at least five years.[5] On the previous page he had given an example of a shift in an indifference map. The individual at first spent a certain amount and 'after a hundred years he will spend' another amount. He remarked that 'men's tastes are very tenacious', but still it seems a curious remark when one considers the normal length of a lifetime, and especially since on yet one page previously he had said that an individual was not exactly the same from day to day. Did the exaggeration indicate that the man who had such penetrating sociological insights was a little uneasy about this aspect of his economic theory?

All these complications were forgotten in the Mathematical Appendix. The essential information about preferences could be obtained directly from observation. He illustrated the method. If one found combinations of goods $(x, y, z,...)$, $(x + \Delta_1 x, y + \Delta y, z,...)$ and $(x + \Delta_2 x, y, z + \Delta z,...)$ among which choice 'is a matter of indifference', then all one needed to know were the limits of the ratios $\Delta_1 x / \Delta y$, $\Delta_2 x / \Delta z$,.... If these were known one could draw up the equations needed for the theory of economic equilibrium. (For a succession of indifference curves one also needed to assume that more is preferred to less. Pareto took this for granted.) Moreover, according to Pareto, ophelimity would not enter such equations: 'Hence the entire theory of economic equilibrium is independent of the notions of (economic) utility, of value in use, or of ophelimity.'[6] In a footnote, he remarked with evident satisfaction that this was where his economics departed from that of the Austrian school. However, that is of less interest here than the question of the invariance of preference orderings. When in the text he gave advance notice of the conclusions he would reach in the appendix, he said: 'The theory of economic science thus acquires the rigor of rational mechanics: it deduces its results from experience, without bringing in any metaphysical entity.'[7] However, in order to remove the metaphysical entity of a quantity of satisfaction, it seems that Pareto, despite all his invective against 'literary economists and metaphysicians', had to introduce an equally chimerical entity. The empirical approach he suggested in the appendix was not an inductive method designed to isolate repeated or normal action from other action: it simply proceeded on the assumption that one would not be recording fortuitous historical facts, but measuring parts of lasting structures. He posited (by implication), the idea that human action was merely the outward manifestation of some deeper, underlying entity. But there was no evidence for this in the bare facts of a recorded indifference among combinations of goods. It was an assumption that should have made a good positivist like Pareto shudder and wince. However, he needed it as the rationale of his avowedly empirical approach.

Let us consider an example from more recent times. Kuenne describes the operational view of consumer's choice in which utility has been eliminated just as the concept of the ether was eliminated from physics.[8] Here the consumer is not seen 'to be consulting his preference function to attain some optimum' but rather the process of maximizing 'is a convention adopted for convenience by the economist'. He so constructs the preference function that it is at a maximum when the consumer does whatever he does. It is therefore 'a repository of observed behaviour, or of response to questions, or of these and assumptions such as transitivity which economize on the direct evidence needed to construct the function'. The purpose of models of consumer choice based on the operational view is, therefore, the definition of choices and not their prediction. Now, in so far as this is a retrospective view used to surmise a consumer's tastes, even von Mises might have agreed with the operational view. But the consistency assumption is also involved. Kuenne went on to say that models of consumer choice do predict (i) 'to the extent that assumptions were used to fill gaps in the construction of preference functions' and (ii) 'in the *trivial* sense that the consumer does what he said he would do or did in the past' (italics added). The difference between the two is apparently that in (ii) the model merely throws back the information that was put into it, while if transitivity in the sense of logical consistency is used in (i) to fill in gaps, one is able to deduce a new choice that was not actually put into the model as a datum. However, even this supposedly very limited predictive capacity of such a model seems to rest on a fundamental assumption that is simply taken for granted, namely, that one can speak of gaps at all in this context and that one can regard the fact that an action is repeated as trivial. Surely, the assumption that people will do what they did in the past, or even what they said they would do, does not fall into the category of trivialities. Is it not more appropriate to put it into the economic category of the heroic assumption?

5.6 Consistent preferences in ordinary discourse

One would have to be a free-will fanatic to deny the meaningfulness of such everyday notions as convention, custom, fashion, habit or personal tastes. I was not trying to suggest in the previous section that there is not a sense in which one may reasonably speak of consistent choice, or that one may not single out as facts that people consistently prefer, at least for some time, certain things to certain other things. The everyday notions of tastes and habits, however, differ in an essential respect from the concept of a consistent preference ordering as used in the theories of consumer choice and of demand. I shall consider this difference in the present section.

One would not find it odd to be told that person X prefers the compositions of Bach to those of Beethoven, and those of Beethoven to those of Brahms (and therefore those of Bach to those of Brahms), nor would one find it odd, on meeting X five years later, to discover that X's tastes in music have not changed. Our everyday knowledge of the tastes of others is so common that we probably do not realize that inductions of a high degree of sophistication are involved. In the case of person X's preferences in music, for instance, we have to be able to recognize the situations in which X is enjoying music, and distinguish them from ones in which music is merely incidental or in which music is a means to an end rather than an end in itself. We would not be surprised to find that X prefers a record of the latest in popular music when entertaining friends, nor would we be surprised to find that X is playing a recording of Wagner's *Parsifal* at full blast on a Sunday afternoon if we suspect that X wants to annoy his neighbour. Again, we would not revise our ideas about X's preferences if we found that the person spent a vacation day climbing a mountain instead of listening to Bach. We may also know that X prefers mountain-climbing to sun-bathing next to a pool, but our knowledge of X's preferences in music and in outdoor recreation does not tell us how the person will allocate scarce recreation time among various activities. Person X may allocate no time to either Bach or mountain-climbing and his or her preference between indoor and outdoor recreation may show no consistency whatsoever. Only rarely do we know of activities that would consistently be preferred to all or even most other activities. If person X's house is on fire (and X is in it) we can predict confidently that he or she will give little thought to either Bach or mountain-climbing. In short, our knowledge of individual preferences involves highly sophisticated inductions, and the facts we arrive at do not constitute an integrated system of preferences that allows us to predict action under all circumstances.

The preference hypothesis in economic theory does not rule out the possibility of common elements among individual preferences, that is, it does not rule out fashions and customs, but these are very seldom the subject matter of analysis at least in the more formal cart of theory. Yet for a subject concerned with aggregates or at least with the simultaneous actions of a large number of people, fashion and custom would seem to be a more manageable subject matter than individual tastes, if these are to be seen in an *ex ante* context. What I said about individual preferences also applies to fashion and custom. If, as Pareto said, Italians prefer coffee to tea we would not be surprised to find that they drink wine also. The type of circumstances in which coffee and wine are imbided may be different. We may also know something about Italian wine-drinking habits. Perhaps they prefer to drink white wine, perhaps even French to Italian wines or large quantities of vin ordinaire to small quantities of expensive wines. But knowing this and that they prefer coffee to tea still does not tell us what proportions of their income, in aggregate, Italians allocate to wine and coffee. However, the recognition of fashions and customs involves further difficulties. Even when the conditions or circumstances under which one expects choices to be made according to fashion or custom are taken into account, our statements about choice (or our *ex ante* facts about choice) do not apply without exception. One may happen to meet a person who has tea in the afternoon and orange juice with dinner, but whom one nevertheless judges to be Italian by various other criteria. The conditions for customary choice may be present (for example, the person is not drinking tea because he or she is entertaining an English visitor, or orange juice because the person has a cold or is a health fanatic) and yet one would not necessarily reject the person's status as an Italian nor the generalization that Italians prefer coffee and wine under certain circumstances.

The problem of perceiving habitual, fashionable and customary action is no doubt the same as Marshall's problem of what is normal action. I think that one must accept that the everyday notion of something normal in human affairs is not spurious and illusory; people have not deluded themselves for countless centuries when they have used it as a guide to action. Marshall appears to have been quite content simply to invoke the everyday notion. This seems unsatisfactory; it is surely desirable that economists should have a more rigorous version of the vaguely and loosely applied, even though sophisticated, inductive methods that may suffice for everyday purposes. That, it seems to me, is how sciences arise out of ordinary discourse. I hope to be able to contribute to something of this nature on another occasion. It is beyond the scope of the study. For present purposes it will be necessary to extend only slightly the cursory glance I have been throwing in that direction.

The association of person X with the music of Bach and the association of Italians with coffee may be seen to be what have been called *ex ante* facts in this study – whether they are structural or causal facts depends only on the way they are established. The fact that these associations are expressed in the form of preferences means that they hold only when certain conditions are met, for example, when the choice is between coffee and tea or when coffee is chosen under circumstances in which others are known to choose tea. An inspection of the definitions of structural and causal facts (Chapter 2, section 2.8) shows that the conditions necessary for an association were there expressed as the set of elements A of a situation identified by defining procedures B. The elements A are of course most important for the cognition of an *ex ante* fact. In the case of a causal fact, for instance, the outcome D, depends as much on the elements A as on the operation X which is performed.

The amount of detail in the specification of the elements A may vary from case to case, but there must nevertheless be a limit to such detail, that is, the set of elements A must be finite. This needs a few words of explanation. I have stressed that it is in the nature of an *ex ante* fact that it must hold in a variety of situations. More generally one may say that it is the purpose of all inductive methods to establish regularities or constant relations which, by the very meaning of the terms, must hold in a variety of situations. This must be so since every situation is likely to be unique in some respect so that there would be no chance that the conditions for the cognition of regularities, that is, the set of elements A in the definitions of *ex ante* facts, would be found repeatedly unless they were abstractions. Such repetition is obviously necessary if we are to speak of regularities or constant relations. The set of elements A must be specified in such a way, therefore, that it is distinguishable from the situations in which it occurs. Since, as I have said, it is unlikely that the description of a situation can ever be exhaustive, this really means that the set of elements A must be finite. The less elaborate the specification of the elements A, the more likely it is, generally speaking, that these elements will be found repeatedly. A fairly frequent recurrence of specified elements of situations is of course a prerequisite for the establishment of *ex ante* facts by induction, and it is also necessary if such *ex ante* facts are to be useful as guides to action.

It may be that the physical sciences have been able to establish many *ex ante* facts in which the set of elements A is a sufficient condition for the outcome D. Experience in the social sciences seems to suggest that few *ex ante* facts in terms of sufficient and necessary conditions may be found and that one must be content with ones in which the sets A and D are what I called potential conditions for each other (Chapter 2, section 2.8, explanatory note 6). The classification of a person as Italian is a potential condition for a preference for coffee: a person's preference for coffee is a potential rendition for being classified as Italian, but of course it is also a potential condition for being classified in many other ways. Induction in the social sciences may therefore be far more difficult than in the physical sciences. The art of induction is to find a useful degree of specificity in the definitions of the sets A and D; the definitions must be specific enough to be useful, yet not so specific that they are hardly ever applicable to real situations. The method of positing hypotheses which is widely used for these purposes is, in my opinion, subject to the qualifications I expressed in Chapter 2, section 2.9. In the case of preferences in an *ex ante* context, these qualifications require us to realize that the only reason we have for believing that useful specifications for the sets A and D may be formulated is the belief that the everyday notions of tastes, habits, fashions, customs, and so on are not illusory.

The important point for present purposes is that these everyday notions merely give us some assurance that we may be able to specify sets of elements A and D in such a way that *ex ante* facts relating to preferences may be found in *certain* isolated situations. These notions do not give us any indication whatsoever that *every* situation that may arise will *necessarily* contain elements for which *ex ante* facts may be stated. Yet it is this quite unwarranted extension of the everyday notion of tastes that seems to have been made for the assumption of consistent preferences in deterministic economic models. Let us look at this more closely. I argued in Chapter 2, section 2.4 that deterministic economic models try to incorporate both the forward-looking view we have when we are consciously aiming at ends and also the passive view we have as observers of the actions of others. Tastes in an *ex ante* context belong in the first of these views, but since these models are not concerned with the achievement of a particular aim by one individual, but with the prediction of the common outcome of the actions and interactions of many individuals as seen by a passive observer, knowledge of isolated *ex ante* tastes applicable in only certain situations will not do. A deterministic model requires the extension of the everyday notion of tastes into a comprehensive system of preferences that has no gaps and according to which every individual consistently ranks all goods that one may come across into an order of preference.

In planning one's action an individual may conceivably find it useful to know that Italians drink coffee and wine under certain circumstances (or that they 'tend' to do this, as potential conditions are often expressed in economics), certain relative prices may be among the conditions that have to be met for this knowledge to be applicable. If relative prices change so that these conditions no longer hold, the individual simply has to try to achieve the aim in some other way or may adopt another aim, that is, choose to do something else. One does not have this latitude in deterministic models, for there we are not taking the view of an acting individual but that of a passive observer. The aims of acting individuals, as well as the tastes and customs of others that these individuals assess and use as guides to action, are all part of an underlying structure made up of the comprehensive systems of preferences of individuals and this underlying structure has no gaps. Price, income and cross elasticities of demand are assumed to exist for every conceivable good so that the choices and actions of all individuals are determined in all conceivable situations. There is, however, nothing in our ordinary experience of life to suggest that such comprehensive systems of preferences exist, and much to suggest that they do not. Nor, until the last 80 years or so, do many people ever appear to have thought that they have existed: it is even doubtful whether many economists really believe themselves and their fellow humans to be so equipped. In this respect, economists have not let their subject grow out of the ordinary experience of life, but have forced it into the mould of classical mechanics.

It appears that it is not always fully realized that the system of preferences that feature in current micro-economic theory are not the tastes that we speak of in ordinary discourse. This has been so even when the consistency of preferences has not been accepted uncritically. In his excellent book on the history of economic thought, Mark Blaug makes some discerning remarks about objections to demand theory based upon 'the inherent instability of wants', objections which he says, 'cannot be lightly dismissed'.[9] Among other things, he says: 'It is clear that consistency means constant tastes and that inconsistency can be interpreted as a change in tastes. Indeed, the "consistency postulate" amounts to the proposition that a utility function exists.' In support of the consistency postulate, he says: 'If the pattern of wants is never stable even for a short period of time, it is difficult to see why business men spend so much money creating wants; why generate new wants if their inherent instability makes it impossible to guarantee that they can be exploited for a definite period of time?'

However, it is surely not the *pattern* of wants that a business person is concerned with. Business wants consumers to prefer its product to those of rivals or to participate in activities for which it caters rather than in certain other activities, and this is quite compatible with the everyday meaning of tastes and preferences. Advertising does not require the existence of gapless systems of individual preferences or patterns of wants. An advertiser wants consumers to prefer a_1 to a_2, a_3..., another wants consumers to prefer b_1 to b_2, b_3..., and so on. Even though all goods compete for the consumer's 'dollar', no one business person expects a consumer to spend his or her entire income on the business's product, and the business person surely does not care how that income is spent as long as a reasonable portion is spent on its product. In other words, business people do not actually contemplate establishing consumers' definite and consistent preference orderings also for a_1, b_1, c_1...., and definite marginal rates of substitutes, income-compensation lines, and so on. It may be that the combined effect of advertising and sales promotion is to create a common life-style or to change an existing one, that is, to affect the 'culture' of a people or the state of activities in Marshall's sense. But individual advertisers do not aim at this and it is not necessary for the success of an advertising campaign. The mere fact that business people promote their sales does not indicate that there is a common life-style and certainly not that there are individual ordinal preference fields.

I shall briefly consider a life-style hypothesis in the final chapter. Here it should be noted that it differs in important respects from the hypothesis of *independent* and consistent ordinal preference fields. A life-style (in the sense indicated here) is an extension of the ordinary notions of fashion and custom, which I considered in this section, into a more comprehensive notion of a fashionable or customary way of life. It is a system of shared preferences, does not cover every aspect of life and leaves room for individual improvisation. Moreover, as I indicated in this section and suggested in relation to Marshall's activities, the preferences involved in a customary way of life may easily be fitted into an *ex ante* context. I want to suggest that the concept of an ordinal preference field without gaps, on the other hand, is derived from an *ex post* context. The rest of this chapter will be devoted to this suggestion. I shall begin with an outline sketch of how the concept of an ordinal preference field appears to have arisen.

5.7 Ordinal preference fields

In classical political economy and in Marx's system, use value and exchange value had remained largely unreconciled. One of the major incentives for the introduction of demand, analysed on marginal principles, into the mainstream of economic theory was the demonstration that exchange value could be derived from use value or utility. This demonstration leads to the idea of a consumer equilibrium. In an *ex post* context consumer equilibrium may be taken as a formal representation of an explanation of how an individual judged, at some moment in the past, the relative usefulness or desirability to him of the last increments of various goods, when the availability of the goods is taken into account. In an *ex ante* context consumer equilibrium seems to imply that an individual's tastes should be conceived as a comprehensive system of preferences consistent over time. It is the latter conception that interests us here.

Possibly in analogy to the old stalwart that more of a thing is preferred to less, the utility a person derived from a good was at first thought to depend only on the quantity of that good, subject of course to the essential condition that utility increased at a diminishing rate. This gave rise to the additive utility function with the form $U=f(a)+g(b)+h(c)+\dots$ (where U is total utility and $a,b,c,\dots$ are quantities of different goods). However, while the amount of utility derivable from a particular quantity of a good was independent of the nature and quantities of the other goods purchased by the individual, the quantities of the various goods bought were of course not independent of each other. The individual so adjusted the various quantities that the ratio of marginal utility to price was the same for each good when its entire income was spent. Therefore to each level of a consumer's income and to each set of prices there corresponded a definite collection of goods that the individual would buy. Once income and prices were given, such a collection of goods would therefore be determined by the individual's tastes. Since the marginal utility of the 'j'th good (a quantity) depended only on the quantity of the 'j'th good and nothing else, it would have been possible to describe a person's tastes fully in a table of marginal utilities. One can list various goods from 1 to n horizontally, and incremental units of these goods from 1 to m vertically. If 'u' stands for the utility the individual derives from an incremental unit of good, the first column then lists the figures for u_{11} to u_{m1}, the first row the figures for u_{11} to u_{1n}, and so on. In this way, u_{ij} is the utility the individual derives from the 'i'th unit of the 'j'th good. Since all the u_{ij} are assessed by the individual, so to say in the same head, the matrix $[u_{ij}]$ describes the individual's tastes or system of preferences. If the table includes all goods the individual can come across and if these goods are available in discrete units only (alternatively, if the columns in the table are replaced by continuous functions), the system of references has no gaps, that is, it is a comprehensive system of preferences.

The additive utility function did not remain unchallenged for long. In 1881 Edgeworth wrote the total utility function in the generalized form $U=f(a, b, c,...)$ instead of $U=f(a)+g(b)+h(c)...$ (with U and a, b, c,... as before) but he did not make a great deal of fuss about the innovation. In 1892 Fisher showed that while one could easily find a principle for measuring utility if utilities were independent (or additive), this was not so in the case of generalized total utility function. The latter could not in general be derived from indifference curves, and furthermore this did not seem necessary for explaining consumer behaviour. Pareto came to similar conclusions and discussed various types of dependence among goods at some length in the *Manuale*. Not only was there the question of complementarity ('soup without salt is not very agreeable, and clothes without buttons are most inconvenient') but also that of substitution when some goods are regarded as inferior to others:

> One who has nothing else eats a lot of corn meal; if he has some bread, he will eat less corn: if he has some meat he will also decrease his consumption of bread. We cannot say what pleasure someone gets from a certain quantity of corn meal if we do not know what other foods he possesses.

In 1915 Slutsky suggested ways of making the analysis of consumer behaviour independent of what he called psychological assumptions and philosophical hypotheses. He cast doubt on the general validity of diminishing marginal utility and proposed that the hypothesis of the additive utility functions should be tested empirically.[10]

The recognition of a dependence among goods and therefore of the inadequacy of the additive utility function fitted in well with the attempts to remove utility from demand theory altogether. In the revised analysis of consumer behaviour, cardinal utility and later any overt reference to utility became superfluous, at least for the purposes of a deterministic model. We saw that an individual's system of preferences can be represented by a table of marginal utilities, u_{ij}, if the total utility function is additive. If, however, there is dependence among goods so that the marginal utility of the 'j'th good depends not only on the quantity of the 'j'th good but also on the nature and quantities of the n-1 other goods chosen by the individual, this method of depicting tastes breaks down. The u_{ij} are no longer unique but vary according to the composition of the bundle of goods chosen. The table of u_{ij} can therefore no longer be said to represent the preferences which determine the composition of the bundle of goods an individual chooses, given income and prices, because the u_{ij} in the table are themselves determined by the bundle of goods chosen.

In order to describe an individual's system of preferences one now needed a list that set out the total utility the individual could derive from each entire bundle of goods that was available to him and that might become available to him if prices or his income should change. Obviously, the individual would choose that bundle of goods from among all those available to him which would give him the highest utility. To be able to identify the bundle with the highest utility, one clearly did not need a cardinal utility; the number of units of utility provided by different bundles was superfluous information. One could now represent an individual's preferences by a system of indices which related each bundle of goods considered to each other such bundle ordinally on the basis of more, same or less utility, or of preferred, indifferent or neither preferred nor indifferent. In two dimensions such a representation took the form of the familiar indifference curves, the convexity to the origin of which ensured that only one bundle of goods was the most preferred for any set of income and prices.

The number of possible combinations of goods (if goods are understood in their everyday meaning) is of course vast. It naturally seems rather far-fetched to assume that one could know the position of each one of these possible combinations in an order of preference, or even that the individual itself could be aware of a complete order of preference covering all possible combinations. In subsequent developments of the theory of consumer behaviour such an assumption was held to be unnecessary, just as the assumption of cardinal utility was unnecessary. Whatever combination of goods an individual chooses could be defined as the most preferred combination of all those that are available to the individual (or as one of a set of most preferred combinations between any two of which the individual is indifferent). An individual's choice therefore reveals something about his or her preferences to the observer and even to the choosing individual, but only about those preferences that relate to the combinations of goods that are actually available when the choice is made, and not about preferences that relate to all conceivable combinations of goods. We saw in Chapter 4, section 4.1 that the assumption that there is a most preferred combination (or combinations) of goods is merely the assumption that an individual acts purposefully. When we interpret behaviour as purposeful action (Chapter 4, sections 4.4 and 4.6) the beliefs and preferences that are revealed (or, more accurately, surmised) may simply be ones that an individual happened to have at a particular moment, without any significance beyond that moment. However, that was not the view taken in the theory of consumer behaviour; the notion of a consistent and comprehensive system of preferences which was inherent in the scheme of cardinal utility was not abandoned. An individual did not only act purposefully but also rationally in the consistency sense that I examined in Chapter 4, section 4.2.

Slutsky had already based his analysis of consumer behaviour on very general assumptions that reflected this viewpoint. He wrote the total utility function in the generalized form and subjected it to the restrictions that (i) the function and its derivatives of the first two orders are continuous and (ii) the function does not change during the period considered. He remarked that both assumed restrictions would probably find approximate empirical confirmation if a group rather than an individual were considered. The remark was put between brackets and, considering the rigour of his analysis otherwise, one may take it that it was meant as an aside.[11] When, 23 years later, Samuelson tried 'to develop the theory of consumer's behaviour freed from any vestigial traces of the utility concept' he started with three postulates. The first amounted to the Slutsky restrictions already mentioned, but continuity was assumed merely for 'mathematical convenience'. (Though continuity may not be essential, I take it that a preference field must be without gaps in the sense that no situation could arise in which an individual would have no preferences.) The second postulate was that all functions were homogeneous of degree zero and the third that choice was logically consistent, that is, if A is chosen rather than B, B must not 'at the same time' be chosen rather than A. A few months later he published a note in which he pointed out that only the third of these postulates was required since the first two were implied by it. Some years later, in a paper in which he gave a verbal and diagrammatic explanation of the integrability problem, he related how, as a result of a remark made by Haberler, it suddenly came upon him 'that we could dispense with almost all notions of utility: starting from a few logical axioms of demand consistency, I could derive the whole of the valid utility analysis as corollaries'. The fundamental axiom – the third postulate mentioned – he now called the Weak Axiom of Consumer Behaviour. Phrased in revealed preference terms, it now appeared to have a time element as well. If A is revealed to be better than B, the basic postulate is 'that B is never to reveal itself to be also "better than" A'. Hicks was to call this two-term consistency. The integrability problem also required transitivity involving more than two terms. Samuelson called it the strong axiom, namely, if A reveals itself to be better than B, B than C, C than D and so on to Z, then 'Z must never also be revealed to be better than A'.[12]

The ordinal preference fields which were now ascribed to individuals were very different from the tastes which we may ascribe to individuals in ordinary discourse. Their main characteristics were that they were consistent and comprehensive or all-embracing and that they satisfied curvature conditions sufficient to ensure that one and only one combination of goods would be chosen under any one set of income and price constraints. For the ordinary notion of tastes, consistency of choice under certain circumstances is a criterion for deciding whether among all the choices an individual makes there are some which reflect what we call the individual's tastes. If one supposed all choices to reflect tastes, consistency as an inductive criterion clearly would not be necessary. However, this supposition is not made in ordinary discourse; it was first made in utility analysis. One of the main purposes of utility analysis was to show that market phenomena, or more specifically exchange value, could be *traced back* to individual choice and judgements of usefulness. The analysis was all-embracing; *all* market phenomena could be so traced back. When utility analysis was seen in an *ex ante* context, choice was associated with the ordinary notions of needs and tastes and therefore with consistency. When utility was removed from this analysis because it was regarded as superfluous or even 'unscientific', what was left was the concept of a comprehensive and consistent ordinal preference field. Consistency was no longer an inductive criterion, but a presupposition about empirical fact – given the same circumstances (only prices and income were considered) the same choice would always be made. The assumption of consistency also meant that preference fields had no gaps. The comprehensiveness of explanation that we are used to in an *ex post* context, in which any and all behaviour can be interpreted as purposeful action, could now also be introduced into an *ex ante* context. An individual's choice did not reveal merely aims and preferences at the time the individual made the choice – the temptations of the moment as Croce put it – but rather it revealed one point in a comprehensive and consistent but otherwise initially unknown ordinal preference field.

It would be incorrect to say that the existence of consistent ordinal preference fields was simply taken for granted. A decade after he first wrote about the consistency postulate, Samuelson seemed to evince some doubt when he said that 'the individual guinea-pig, by his market behaviour, reveals his preference pattern – if there is such a consistent pattern'. He may well have meant that the guinea-pig's recorded behaviour may give rise to non-integrable equations. The integrability problem appears to be about whether ordinal preference fields exist, but the question of existence is here different from the one I have been investigating. One must distinguish between logical consistency and consistency over time. I argued in Chapter 4 that this distinction is in fact difficult to make in an empirical investigation, but logical consistency has of course a separate meaning in axiomatic constructs because the temporal order of experience does not enter the picture. Whether a differential equation is integrable is a purely mathematical question. When it is applied to preferences, it can surely not be interpreted to relate to anything other than logical consistency, or rather transitivity, because time is not an issue in the mathematical question. (The path of integration, as the discussion of Pareto's handling of integrability seems to have shown, has nothing to do with the temporal order of experience.) Non-integrability in relation to preferences must therefore be understood as logically inconsistent choice, with the question of the duration of preferences left open. In the paper on the integrability problem from which I quoted earlier, Samuelson argued in favour of the integrability hypothesis by asking why someone without an integrable preference field should be thought of as acting irrationally. Why should the person not be thought of as 'changing his demand behaviour constantly and capriciously ... why should his demand have any time invariance?' However, he immediately mentioned two counter-arguments that could be brought against him. One was that 'a man might display consistent demand behaviour through habit or crude rules-of-thumb not consistent with an ordinal preference field'. The argument, however, seems to be based on the presumption that the people's habits and norms make up a comprehensive albeit inconsistent system covering every choice a person ever makes. This seems to be most unlikely.[13]

Hicks regarded integrability as a question without importance to economics.[14] When therefore he examined the Samuelson-type consistency tests, by which one could disprove that individuals act according to ordinal preference fields, he presumably had a temporal context in mind. As an empirical procedure he did not rate them highly. They were 'essentially tests of individual behaviour' and the hypothesis that 'the Mr. Brown or Mr. Jones who lives round the corner' has a consistent scale of preferences 'does not deserve a moment's consideration'. When Slutsky's suggestion is followed up and the tests are applied to groups, all kinds of difficulties arise. Hicks concluded 'that there is in practice no direct test of the preference hypothesis'. Its status, rather, is that of an instrument for arranging empirical data 'in meaningful ways'. It is not the only hypothesis that could be invented for this purpose, but 'one which, initially at least, seems to be the most sensible hypothesis to try'.[15]

5.8 Complementarity and similarity

5.8.1 The definition of a good

The additive utility function could not account for complementary goods. This was perhaps the most cogent reason for the adoption of the generalized utility function, which in turn facilitated the rejection of cardinal utility and the introduction of the concept of ordinal preference fields. The implications of complementarity merit closer attention. Let us return to what Pareto said on the subject. The ophelimity one derives from coffee is considerably enhanced if one also has a cup. It is further enhanced if one also has sugar and still further, though presumably not as much, if one has a spoon. The usefulness or ophelimity of these four items together is therefore not the sum of their ophelimities separately. One may overcome any problems that may arise from this by regarding coffee, cup, sugar and spoon as one composite commodity. But, Pareto warned, one thereby creates a greater problem than one solves. Where does one draw the line? Should one not regard every possible combination of goods as a composite commodity? We would then 'multiply the number of goods without bound'.[16] Pareto did not mention that this is in principle what the generalized utility function does.

The demarcation of goods has not been treated as fully in economics as perhaps it should have been. Pareto, like Marshall, took the line that the definition of a good had to suit the context of a particular problem. This is no doubt sensible, but it does not solve the problem of how goods are to be defined in more formal analysis. Arrow and Hahn, in the work from which I have quoted before, said the following: 'A good may be defined by its physical characteristics, its location in space, and the date of its delivery. Goods differing in any of these characteristics will be regarded, as different.'[17] If time and space are understood in the normal way, this means that each good is unique. There need be nothing strange in this. There is a sense in which, for instance, every residential house is unique. Two houses may have the same design, and the bricks, mortar, and so on may have the same chemical composition, but they are nevertheless quite different if one is built away from others on a ridge with a magnificent view of mountains and the other is squeezed between two warehouses in a smoky and dirty environment. The title deeds may specify only that a house and its surrounding garden is owned, but the houses the owners enjoy cannot be separated from their respective environments. The appreciation of a situation in its entirety is not the only nor even the most common case to which the complementarity argument applies. It applies also when goods are chosen not as ends in themselves but as part of the means towards some more long-term goal.

The usefulness of goods can then not be abstracted from a situation as a whole, for the goods are meant to combine in a certain way with the other elements of the situation to advance the buying individual towards his or her long-term goal. More of this later.

If the argument by which the additive utility function is rejected is taken to its logical conclusions, then what is indicated by the dots in U=f(a,b,c,…) must be extended to include a whole situation, in all its particularity. Since it seems impossible that any two situations could ever be exactly alike (even if only because time is unidirectional), all that the generalized total utility function would then tell us is that every situation in which an individual may find himself or herself has a certain desirability to him or her. Further, we could use the function in an explanation of an individual's action, namely, that the individual selects from among all those situations that it expects to be able to bring about by its own choice or action, that one which is most desirable or most prefered. Now this is very much like the view we take when we see action in an *ex post* context – when we interpret the already completed actions of others.

However, deterministic models are seldom intended to be devices solely for explaining the past. In an *ex ante* context complementarity raises problems. One can of course simply postulate that there exist comprehensive systems of preferences each of which assigns a relative position, if not to each unique situation, then at least to each unique good. Perhaps that is how Arrow and Hahn conceived the matter. However, their definition of a good was not designed for a revealed preference approach. It is quite obvious that that approach requires a different definition. In a purely technical sense the revealed preference approach could of course deal with unique situations. If the commodity space in which the ordinal preference field is to be revealed is given sufficient dimensions, one can simulate unique situations as closely as one likes. The space assigned to a person at birth would be blank and after three score years and ten it would contain a large number of points (possibly quite isolated, in view of the vastness of the space) giving the outline of an ordinal preference field. However, the exercise would serve very little purpose, except perhaps as a record of a person's life. The assumption made about the field, namely, that preferences are consistent, would be quite superfluous since (unless one believes that history repeats itself exactly) the situations chosen in the past, and the possible situations rejected, would never be available again, that is, choice would always be between situations that have never been available before so that there could be no question of consistency.

Clearly, the revealed preference approach was intended to be used with a definition of goods that makes it possible for combinations of goods to be available repeatedly. In fact, the only limitations on the availability of all conceivable combinations of goods in that view are the prices of goods and individual budgets. The full implications of complementarity and of the generalized utility function for the consistency of choice are not so obvious when an individual is seen to be choosing only between x's, y's and z's, the respective units of which are all perfectly homogeneous.

When one compares the Arrow-Hahn definition of a good with the everyday notion of goods, it would appear that the former in some respects contains more and in others less than the latter. Let us take coffee as an example of a good. It seems reasonable to suppose that one of the criteria for classifying a substance as coffee is the effect the substance has on the palate and one may accept that this effect is susceptible of explanation in terms of 'physical characteristics'. But the everyday notion of goods does not seem to need the space and time dimensions. Coffee beans in different places and delivered at different times are no doubt different things but one would normally say they are the same good if they have appropriate physical characteristics. These, furthermore, do not have to be identical; a certain similarity suffices. On the other hand, the Arrow-Hahn definition does not include a very common criterion for classifying things as one or other good, namely, the purpose for which a thing is normally used. Two houses may be quite unlike in design, physical characteristics, location and date of completion and yet we speak of them as houses because of a similarity in the purposes for which we expect them to be used. One could say the same about the classification of things as food, clothing, furniture, and so on. Again, the purposes in question do not have to be identical – it may be foolish to speak of identical purposes – they merely have to be similar. The idea of similar purpose is of course behind the concept of substitution.

Similarity is a very troublesome idea, but one could hardly deny that it is meaningful or that it is behind the everyday notion of goods or commodities. The task of drawing up a definition of a good, that is, of drawing a line between similar and dissimilar, is immensely difficult. One may consider Pareto's example of inferior goods and the corresponding dependence among goods (section 5.7). Corn meal is considered inferior to bread, bread to meat. The desirability of corn meal to an individual depends on the amount of bread and meat the person possesses. Now, one may regard corn meal, bread and meat as distinct goods, that is, as dissimilar. But then, if there are two situations in which the (historically) same individual has or buys a loaf of bread, one may not, on Pareto's argument, regard them as similar situations without also taking into account the amount of corn meal and bread held by the individual. On the other hand, one may regard the three as similar and speak only of food. But then, if there are two situations in which an individual has the same quantity (by weight, volume, calories?) of food, one may not regard them as similar situations without also taking into account what kind of food it is, for, by Pareto's argument, the individual's attitude to the food would not be the same if in one case it consists mainly of meat and in the other mainly of corn meal. Corn meal, bread and meat are neither sufficiently dissimilar to be regarded as distinct goods nor sufficiently similar to be regarded as the same good. I shall not enquire further into the matter here. The point at issue is that satisfactory solutions to such problems have to be found before one can speak sensibly about consistent choice.

5.8.2 A formal analysis of consistency

This general discussion of complementarity and similarity has prepared the way for a final comparison between consistent choice in the everyday sense and the consistency of preference fields. First, let us recall that the everyday notion of tastes, that is, of consistent choice, has the form of an *ex ante* fact. When we say that person X has a preference or a taste for whisky we mean that in certain circumstances X may be expected to choose whisky rather than some other beverage. It should be clear by now that *ex ante* facts also require definitions based on similarity. It must be possible to regard certain quantities of liquid found in different places, at different times and in differently shaped bottles with different labels all as sufficiently similar to be called whisky and sufficiently dissimilar to other liquors to be regarded as a distinct good. Further, the circumstances in which one expects X to manifest his or her preference (for example, being offered a drink) must also be defined on the basis of similarity. They are *types* of circumstances: one would not expect circumstances to repeat themselves in every detail. One may observe person X in a wide variety of circumstances and those in which one expects X to manifest preference for whisky may occur rarely. One would therefore not be surprised to find X drinking water, tea, and so on as well, nor would one expect to be able to explain or predict these other actions using the knowledge of X's preference for whisky. There can therefore be no question of a preference *field* or a comprehensive system of preferences. Person X's preference for whisky quite obviously cannot explain or predict all X's purchases.

Let us now consider the consistency of preference fields. As the concept of consistent choice has been developed in the theory discussed in the previous section, an individual would not have to buy the same amount of each good in each period in order to choose consistently. This would be so only if prices and incomes were constant, in which case an individual with a consistent preference field would be expected to settle into a stationary state. In other cases consistency of choice is tested by comparing all purchases in one period with all purchases in another; a procedure which allows for changes in prices and income and for the existence of complementary goods and inferior substitutes. One may consider Hicks's version of the consistency test here (though it should be remembered that he did not think it worthwhile to apply the test to individual choice nor feasible to apply it to group choice). Let us confine ourselves to two-term consistency and follow his notation.[18]

There are therefore two situations indicated by the suffixes 0 and 1. (Hicks refers to situations and not to periods.) In each situation the individual chooses a collection of goods. These are indicated by q_0 and q_1. There are also sets of prices indicated by p_0 and p_1. The notation $(p_0.q_0)$ is taken to indicate the various quantities of goods chosen in situation 0 multiplied by their respective prices, and so for $(p_1.q_0)$ and so on. The individual's income in situation 0 is therefore also indicated by $(p_0.q_0)$.

Two observations are made and then the consistency test is the following: An individual has chosen *inconsistently* if $(p_0.q_1) \le (p_0.q_0)$ *and* $(p_1.q_0) \le (p_1.q_1)$ unless q_0 and q_1, are identical. (When indifference is allowed for, there is the additional proviso that at least one of the relations must be an inequality.) In other words, if the individual has a consistent preference field, he or she cannot prefer q_0 to q_1 in situation 0 and q_1 to q_0 in situation 1. If only one or neither of the relations holds, the individual has chosen 'not inconsistently'.

'Not inconsistently' does not mean 'consistently'. The reason for this is plain. Suppose the relation for situation 0 in retrospect turns out to be $(p_0.q_1) > (p_0.q_0)$, that is, q_1 would have been too expensive for the individual in situation 0. The test then could not possibly have revealed either inconsistency or consistency of preference for one over the other because both q_0 and q_1 were not available in both situations. Nor could one expect the individual to have chosen q_0 consistently in both situations because the individual would not have spent all its income if it had chosen it in situation 1. Under these conditions choice would be revealed to be 'not inconsistent' because it could not be revealed to be inconsistent and not because it was consistent. The position would be little different if every good was unique, since it can hardly matter to the individual whether q_1 is unavailable because it cannot afford it or because it is not on offer. But then there could be no question of either inconsistent or consistent preference. 'Not inconsistent' can therefore not be taken to mean 'consistent'. 'Not inconsistent' does not really mean anything. The only instance in which 'not inconsistent' would be taken, on repetition, to mean 'consistent', in the sense of an empirical regularity, is that in which q_0 and q_1 are identical. However, there would then not necessarily be a consistent preference field, but only a consistent choice. The conditions necessary for conducting a meaningful test for consistent preference are therefore that q_0 and q_1 are different and are both available in both situations as the consistency test is formulated, an inconsistent preference would necessarily be revealed when these conditions are met. (Unless indifference is allowed for and both relations are equalities.)

Since 'not inconsistent' is uninteresting the best one can do is to investigate the conditions for inconsistency and to compare them with the conditions for consistent choice in the everyday sense. For these purposes it is desirable to have a more formal way of stating similarity and the assumption I have been making throughout this section, namely, that every situation is unique. Let the symbol S denote a situation. By situation I mean circumstances that relate to an actual or imagined here and now and that are in the consciousness of an individual. Situations may include those into which the individual has come by accident, those that he or she has attained (that is, deliberately) through his or her choice or action and those that the individual has imagined or expected he or she could have so attained. Whether the number of S in a person's life is finite is not important here, nor need we be too concerned about the question of where one S ends and another starts. We can therefore ascribe $S_0, S_1, S_2, \ldots$ to an individual, or other situations distinguished from each other by a first subscript. I shall write S'_i when it is thought to have somehow arisen out of S_i. Let us assume that S_i can be described by enumerating its elements. A description consists of an ordered set of any number of arbitrarily selected elements. The elements are either types (generic concepts) or numbers. Thus a description may include, say, one litre of milk or four kilogrammes of sugar, and so on. There may also be complex types such as 'being offered a drink', and so on. The individual's knowledge (*ex post* and *ex ante*) and intentions and expectations, which are situations in themselves, may also be included in a description. It is assumed that a description can *never be exhaustive*. The notation S_{in} stands for a description of S_i, consisting of the ordered set n of particular elements in a particular combination. The first subscript of S therefore denotes a specific situation and the second and any further subscripts a specific description of that situation.

Since the elements of a situation are types and numbers, they may be found in other situations as well. It may therefore be possible to find an S_{im} and an S_{jm} that are identical in the sense that they are sets of the same types and numbers in the same combination. If S_{im} and S_{jm} are identical S_i and S_j are *similar in m*. (They are only similar because their descriptions could have been elaborated to a point where they would no longer be identical.) The assumption that *every situation is unique* may now be stated as follows: It is always possible to find an S_{in} such that there is no S_{jn} that is identical. This assumption, it may be noted, does not depend on an assessment of how deeply people analyse situations. For instance, von Mises's description of action (Chapter 4, section 4.4 and Chapter 2, note 12) could now be put as follows. An individual in S_{ia} tries to reach S'_{ib}, where b is more desirable to him or her than a. Now, it may be, if it is an unthinking individual, that there is an S_{jb} identical with S_{ib}. The uniqueness assumption merely states that the individual could have extended the description into S_{ibz} so that there is no S_{jbz} that is identical with it. (The individual may then no longer find bz more desirable than a.) I do not think that it is an unreasonable assumption.

Let us now apply this analysis to the two types of consistency under investigation. The notion of consistent choice in ordinary discourse has the form of an *ex ante* fact. In its simplest form, once it has been established, such an *ex ante* fact may be described as follows: One has in mind two ideal-type situations S_{1A} and S'_{1D} (where A and D correspond more or less to these letters in the definitions of *ex ante* facts). If one judges an individual to be in S_0 similar to S_1 in A then one may find or cause (depending on whether it is a structural or casual fact) an S'_0 similar to S'_1 in D. The art of induction is to find an A and D by trial and error such that the combinations S_{0A} and S'_{0D}, S_{1A} and S'_{1D}... occur frequently enough for one to formulate the ideal types and to regard them as expressions of an empirical regularity. Consistency is then an inductive criterion and consistent choice derives its meaning from inductive methods. It follows from the uniqueness assumption that S_{0A} and S_{1A} could have been extended until S_{0Aa} and S_{1Ab} are not identical. As I have stressed previously, *ex ante* facts hold in a variety of situations, that is, it is in the nature of inductive abstractions or of *ex ante* facts that the A above should be independent of the a,b,c,... that may also be asserted about the situations in which it is found. In the present context this is the most important point. It is a necessary condition of induction that the A in S_{1Ab} should be independent of b; there should be, so to say, no complementarity between A and b. If there appears to be dependence, and this may be the rule rather than the exception in economics, there can be no consistency because the induction of *ex ante* facts or if consistent choice is not possible. Economists have sometimes, perhaps intuitively, used a corollary of this, namely, if consistent choice cannot be found then there must be complementarity. Some others, as we have seen, have found it more meaningful to say that there is free will.

Let us now consider Hicks's consistency test. Here the notation will become a little more complicated. First, in order to avoid unnecessary suffixes, I shall adapt Hicks's notation and make $q_0 = x$ and $q_1 = y$. There are two situations, S_0 and S_1, and the individual chooses inconsistently if x and y are available in S_0 and in S_1 and prefers x to y in S_0 and y to x in S_1. Let us describe this more fully in terms of choice and action. In contemplating or planning action in S_0, the individual imagines alternative, mutually exclusive situations, S^1_0, S^2_0, S^3_0,..., which the individual expects to be able to attain. We are interested only in those which the individual can attain by making purchases. The consistency test requires that among these there should be two which can be described as

$$S^1_{0x} \text{ and } S^2_{0y} \tag{1}$$

Similarly in S_1 there must be the options

$$S^1_{1x} \text{ and } S^2_{1y} \tag{2}$$

We assume that the cost of x and y is different in S_0 or in S_1 or in both (to exclude the possibility that the individual's choice can be taken to reveal indifference between x and y).

To reveal the inconsistency the individual must be observed to choose

$$S^1_{0x} \text{ and } S^2_{1y} \tag{3}$$

We denote an index of a preference ordering or of an indifference level by I. This symbol is given two subscripts. The first refers to the situation in which, or the time when, a choice was contemplated and the second to a collection of goods. We may therefore write

$$I_{0x} = f_1(S^1_{0x})$$

$$I_{0y} = f_2(S^2_{0y}) \tag{4}$$

$$I_{1x} = f_3(S^1_{1x})$$

$$I_{1y} = f_4(S^2_{1y})$$

It follows from (1), (2), (3) and (4) that

$$I_{0x} > I_{0y}$$

$$\text{and } I_{1y} > I_{1x} \tag{5}$$

The consistency test is apparently based on the argument that the time at which observations are made should not affect the individual's choices if the individual has a consistent preference field, that is, if preferences are consistent over time. The time subscripts may therefore be left out. We then have from (5)

$$I_x > I_y \text{ and } I_y > I_x \tag{6}$$

This obviously reveals an inconsistency, since each index can now be shown to be greater than itself.

However, on the basis of the uniqueness assumption S^1_0 and S^1_1 are merely similar in x and S^2_0 and S^2_1 similar in y. The description of the situations could be extended to a point where each is different from any of the others. One may therefore leave out something of importance when one ignores the time subscripts of the indexes of indifferences levels. Let us consider this. It follows from the assumption that the description of a situation can never be exhaustive that S^1_{0x} and S^2_{0y} cannot be full descriptions of S^1_0 and

S^2_0. On the other hand S^1_0 and S^2_0 are both options that the individual expected to be open to him when he contemplated the matter in S_0. Since we are interested only in the individual's choice between purchasing x and y, we must assume that S^1_0 and S^2_0 are similar in all respects except for the purchase of x in one and y in the other. We may therefore make the descriptions of S^1_0 and S^2_0 as detailed as we like and still say that

$$S^1_0 \text{ and } S^2_0 \text{ are similar in v} \tag{7}$$

On a similar argument we may say that

$$S^1_1 \text{ and } S^2_1 \text{ are similar in w} \tag{8}$$

It is a condition of Hicks's consistency test that x and y should have been available in both S_0 and S_1. To meet this condition y had to be not more expensive than x in S_0 and x not more expensive than y in S_1. If x was cheaper than y in S_1, then the individual would have had some income left over had he or she bought x instead of y. We denote this residual by 'a'. Similarly, we denote the residual in S_0 by 'b'. We may therefore conclude that the individual really had a choice between

$$x \text{ and yb in } S_0$$

$$\text{and xa and y in } S_1 \tag{9}$$

Either 'a' or 'b' can be an empty set (or less pretentiously nothing) but not both, since then the individual's observed choices can be taken to reveal an indifference between x and y (if observation times are ignored) and the verdict of the consistency test must necessarily be 'not inconsistent'.

From (7), (8) and (9) we may now rewrite the descriptions in (1) and (2) as follows:

$$S^1_{0xv} \text{ and } S^2_{0ybv}$$

$$S^1_{1xaw} \text{ and } S^2_{1yw} \tag{10}$$

(It may be noted that no pair of these situations is identical.)

From an inspection of (4) and (10) we can now rewrite the inequalities in (5). However, to simplify matters we can now drop the symbols for situations, since the situations are now described to any desired degree of detail. We can also drop the time subscripts from the I's as in (6) and use the I's to indicate the function giving the values of the index of indifference levels. We then have:

$$I_x(xv)>I_y(ybv)$$

$$I_y(yw)>I_x(xaw) \tag{11}$$

The consistency test made the implicit assumption that more is preferred to less and therefore (since a and b are residual amounts of income) xa is preferred to x and yb to y. Even with this assumption there is only one way in which the observed choices can be taken to reveal inconsistency, and that is if one can rewrite (11) in the form

$$I_x(x)+I_v(v)>I_{yb}(yb)+I_v(v)$$

$$I_y(y)+I_w(w)>I_{xa}(xa)+I_w(w) \tag{12}$$

Then it would follow that

$$I_x>I_{yb}>I_y>I_{xa}>I_x \tag{13}$$

which would reveal an inconsistency. But to reach this conclusion we have to write up the results of observations as in (12), and (12) is in the form of an *additive* utility function with all that this implies about cardinal and even measurable utility. If the results are left in the form of (11) then we do not know whether x is preferred to yb, ybv to yw, y to xa or xaw to xv because we do not know what complementarities there are or may be between purchases and v and w. The results of the consistency test could never be anything other than that choice was not inconsistent; and this, I have argued, does not really mean anything. On similar lines one could argue that one can never observe an individual's indifference between goods, even when questioned after a choice has been made. One can record only one's choice between unique situations.

The question is of course whether the v and w in the above analysis may be ignored – as is done in standard micro-economic theory. Can one confine one's attention in the theory of demand only to the purchases made by individuals and take it that there is no complementarity between x and v or w and between y and v or w, even though in the generalized utility function one recognizes complementarity among the components of x and among the components of y? Perhaps one could argue that demand theory is based on a fine empirical assessment of the threshold of human sensibility, and that this threshold happens to coincide with all purchases made. There is, however, little evidence of such assessments having been made, let alone agreed upon.

I expressed the view that complementarity may be used as a catch-all to explain any lack of consistency in observed choices. But the notion of complementarity was not of course simply invented. It has an introspective basis, and it is only on this basis that one can really consider the question of whether there is complementarity only among the things purchased in the period over which an individual is observed. If a person goes out and buys ten litres of petrol and a packet of dog biscuits, would it be reasonable to say that it must have been this particular combination of goods that tickled his or her fancy and that, while we dare not regard it as anything but a combination of goods for fear of committing the cardinal-utility sin, we may safely ignore anything else the person may have had in mind? The other things the person may have had in mind may have included not only a motor car, a dog and the other things he or she already has or may make use of, but also everything the person happened to know and expect at the time, the hopes, the fears, the ideals and possibly the indigestion. Whether all this can be safely ignored so that the preference function need contain only goods purchased, possibly with a few extra items that are sometimes included such as value of assets, expected income or income of others, really depends on a question economists may not regard as within their province. One would really have to consider at what point the consumption of a good culminates, or fizzles out, in a glow of satisfaction, well-being or whatever. In this and, as common experience seems to suggest, in very many cases, the goods bought surely do not create such a glow; they are means to further ends. The strict conceptual division between household and firm may conjure up a vision of consumers who merely indulge their appetites and are never concerned with means (other than their income). Menger's scheme of goods of the first and higher orders and the more recent suggestions of Becker, which I mentioned in Chapter 3, section 3.5.6, allow a greater latitude. If an individual intends to combine the goods he or she buys in a definite way with the other elements of the situation in order to achieve a more distant goal, the usefulness of goods depends on the other elements, so it may depend on the minutiae of the situation.

Clearly, economists cannot be expected to deal with all the things the economic subject may have in mind. But it is not legitimate to exclude them by inventing human beings with whom we are not familiar: it is far more acceptable to exclude them by applying the blanket term complementarity. If x is chosen on one occasion and y on another and both x and y are always available, we may invoke complementarity and say that something must have been present to make x more useful the one time and something else to make y more useful the other time. If the validity of the uniqueness assumption is accepted, the argument against cardinal utility, namely, that a person's choice of a particular good cannot be considered in isolation, leads to the logical conclusion that a person's choices for combinations of goods in different situations must necessarily be not inconsistent and that the question of whether they are actually consistent simply cannot arise. But if it is not meaningful to ask whether a person's choices for various goods *considered as a whole* were consistent with the person's previous choices, then it hardly seems sensible to assume that they always are, that is, that there are consistent preference fields.

However, there is one special case in which this argument would not apply, and that is where every purchase ever made is in accordance with what in ordinary discourse we refer to as a disposition, habit, fashion, custom, and so on. Let us consider why this is a special case. Consistent choice in the everyday sense has the form of an *ex ante* fact found by induction. Such induction, as we have seen, is possible only when some elements of situations are independent of, or not complementary to, all the other elements. Our everyday experience suggests that it is not impossible to find such independent elements, but also that they are likely to be isolated. (An individual's taste for whisky cannot explain all his or her purchases.) Induction, in fact, consists of juggling definitions around to see whether any elements can be found that appear to be independent. The description S'_{ID} which characterized part of an *ex ante* fact in the foregoing analysis is therefore selected by an inductive criterion. The q_0 and q_1 of the consistency test or the x and y of the foregoing analysis, on the other hand, are not selected by an inductive criterion; the comprehensiveness implied by the term 'preference field' requires them to be *all purchases* made in a situation or during a period. Since the criteria for selection are different there is no reason for supposing that the collections of goods x and y should necessarily coincide with a D in S'_{ID}, that is, for supposing that all purchases are necessarily made in accordance with a disposition, a habit, a fashion or a custom.

It is not logically impossible that they should coincide, but it would be a very special case. It is the case in which a deterministic model is possible because an *ex ante* fact is at hand for every occasion. It should be noted that it must be an entire *combination or collection* of goods that is chosen according to a *single* disposition, habit, fashion or custom. As we saw in section 5.6, the existence of *several* dispositions, and so on, does not tell us in what proportions goods are bought. Italians may prefer coffee to tea and wine to beer, but this does not tell us the combination of coffee and wine bought. A custom would have to specify purchases in detail, and a new custom would have to become operative as soon as conditions (such as prices) change. All this seems to me to indicate how very unlikely the special case is, and that the union of consistent preferences and comprehensive fields in the concept of a preference field is not a very happy one.

I realize that the preference hypothesis was probably not intended to be taken as literally as I have taken it here. After all, Samuelson called it a logical axiom, Hicks an instrument of arrangement that cannot be tested. However, even if that is its present status, it surely has not been posited quite arbitrarily. It has evolved from our everyday understanding of choice. The view that I have been trying to develop in this and in the preceding chapter is that it has evolved from disparate sources, and this, in my view, accounts for the co-existence of consistency and comprehensiveness in the same preference concept. Let us recall that in our explanations of the past we certainly may, and usually do, ascribe aims and preferences to any individual whatever he or she does – if one likes to put it that way, every action of any individual reveals a preference. Explanations of the past can therefore be quite comprehensive. In principle there is no limit to the detail that we could unearth. But the surmised aims, and the preferences they imply, are embedded in an intelligible account of a unique course of events. In this course of events they may appear variously as fleeting whims, as mistakes later regretted or as long-term goals towards which many minor decisions including many of the purchases made were merely the means. In short, they need not be the consistent choices, the tastes, dispositions, fashions or the normal action that in another context we would use as guides to action.

However, if one holds the deterministic belief that explanation of the past and prediction of the future are the same process moving in opposite directions (Chapter 2, section 2.7) one must presume that the type of preference found in each is the same. It must, then, seem natural to equate the preferences one derives from explanations of the past with the tastes and propensities on which one bases one's expectations of the future. Since an individual's every action in the past may be taken to reveal a preference, it must then also seem natural to suppose that an individual has a gapless, comprehensive system of tastes or propensities, that the individual has a consistent preference field. *Ex post* facts, it seems to me, have been mistaken for *ex ante* facts.

Notes

[1] Talcott Parsons, *The Structure of Social Action* (Glencoe: McGraw-Hill, 1937) pp 129–41 and 159–69, made a great deal of Marshall's activities. On his interpretation, Marshall saw progress as the development of economic virtues, namely, rationality, frugality, honourable dealing, enterprise, pride in craftsmanship, the pursuit of excellence for its own sake, a delight in the exercise of mental and other faculties, and so on. To the economist unaccustomed to associating Marshall with such ideas, Parsons, the sociologist, may seem to be reading too much into Marshall. However, one can find much in the 'Principles' to bear out Parsons's interpretation, especially in Book III, chapter II and in the final chapter of the book. In that chapter Marshall distinguishes between the standard of life and the standard of comfort. The latter is what we could now call the standard of living; the standard of life relates to the degree to which values such as those listed above have been cultivated.

[2] Carl Menger, *Grundsätze der Volkswirthschaftslehre* of 1871. See either *The Collected Works of Carl Menger*, vol 1 (London: London School of Economics Reprints, 1934), or *Principles of Economics*, translated and edited by J. Dingwall and B. Hoselitz (Glencoe: The Free Press, 1950). Page references are to the London School of Economics reprint of the original version.

[3] L. von Mises, *Probleme der Wertlehre*, edited by L. von Mises and A. Spiethoff (Leipzig: Duncker, 1931).

4   Oskar Morgenstern, 'Die drei Grundtypen der Theorie des subjektiven Wertes' in *Probleme der Wertlehre* herausgegeben von Mises & Spiethoff (Leipzig: Duncker, 1931) p 8. The problem, of how an individual equates marginal utilities or whatever when he consumes one thing at a time arises only when wants or needs are objectively given to him, as otherwise we simply assume that he comes to equate them and infer wants from this. (In present-day theory this problem is not likely to arise in any case because time has more or less been excluded from consideration.) Morgenstern followed Menger in insisting on an objective basis for Austrian theory as it would otherwise be 'unscientific' (p 10). Austrian analysis was not a subjective theory of value, he said, but a theory of subjective value.

5   V. Pareto, *Manual of Political Economy*, first published 1906, translated from the French edition of 1927 by A. Schwier (London: Macmillan, 1971) pp 188–90.

6   *Ibid.*, pp 393–4.

7   *Ibid.*, p 113.

8   Robert Kuenne, *The Theory of General Economic Equilibrium* (Princeton: Princeton University Press, 1963) pp 54–5.

9   Mark Blaug, *Economic Theory in Retrospect* (2nd edn, London: Heinemann, 1968) p 358.

10 F.Y. Edgeworth, *Mathematical Psychichs* (London: Kegan Paul, 1881) pp 20 and 104. Fisher's treatment of this question in his 'Mathematical Investigations in the Theory of Value and Prices' is reviewed in G. Stigler, 'The Development of Utility Theory' in *Utility Theory: A Book of Readings*, edited A.N. Page (New York: Wiley, 1968), pp 91–4. Pareto, *op. cit.* (note 5) pp 180–8. E.E. Slutsky, 'On the Theory of the Budget of the Consumer' translated by O. Ragus from the *Giornale degli Economisti*, 1915, in *Readings in Price Theory* (London: Allen & Unwin, 1953) pp 27–56.

11 Slutsky, *ibid.*, p 30.

12 P.A. Samuelson, *Collected Scientific Papers of Paul A Samuelson*, vol 1, edited by J.E. Stiglitz (Cambridge, MA: MIT Press, 1966) pp 4–7, 13–14, 89–91.

13 *Ibid.*, pp 64, 81, 92, 94.

14 J.R. Hicks, *Value and Capital* (Oxford: Clarendon, 1939) p 19. 'When more than two goods are being consumed, it is possible that the differential equation of the preference system may not be integrable. This point fascinates mathematicians, but it does not seem to have any economic importance at all, the only problems to which it could conceivable be relevant being much better treated by other methods.'

15 J.R. Hicks, *A Revision of Demand Theory* (Oxford: Clarendon, 1956) pp 17, 55–8, 191.

16 Pareto, *op. cit.* (note 5) p184.

17 K. Arrow and F.H. Hahn, *General Competitive Analysis* (San Francisco and Edinburgh: Holden-Day, Oliver & Boyd, 1971) p 17.

18 Hicks, *op. cit.* (note 15) pp 108–10.

6

The Genetic Understanding
and Institutions

6.1 Explanation and induction

At the end of Chapter 3 I suggested that an equilibrium model in which the preferences of individuals were forever changing has some considerable advantages and I therefore proposed an enquiry into the variability of preference orderings. This enquiry has now been completed. We may draw the following conclusions from it. The notion of consistent choice that finds expression in everyday terms such as a disposition, taste, habit, fashion or custom has the form of an *ex ante* fact. Provided the notion is not illusory, dispositions, tastes, and so on are therefore independent of particular situations and have a certain invariance, though one would expect them to change from time to time. They do not, however, constitute preference fields. The concept of a comprehensive preference field, or a preference ordering that assigns a relative position to every good or combination of goods available to an individual, is based (in the view taken here) on a confusion between *ex post* and *ex ante* facts. It fuses into one concept the consistency of choice in an *ex ante* context and the comprehensiveness of explanation we are used to in our genetic understanding of events. In this view it would be senseless to speak of the consistency or variability of a preference ordering, because there is nothing to vary. It is necessary only to distinguish between consistent choice and choice as an *ex post* fact.

In this final chapter I shall consider some of the implications of this point of view and thereby also attempt to tie up some of the loose ends left in previous chapters. It will, however, not be possible to deal with these implications in anything but broad outline. This chapter will thus be in the nature of a postscript indicating some possible avenues of further investigation.

A few remarks about what is often called descriptive economics may here serve as an introduction. I have in mind empirical studies of, say, a wage dispute in some industry, the marketing of some agricultural product, the development of a particular mining activity, the protection of an industry, and so on. From an epistemological point of view there is often something peculiarly ambiguous about some (and I do not mean all or most) of these studies. The writer's intentions are not always clear. Is the writer giving a narrative account of some unique course of events which is satisfying and meaningful in itself, but which has no more significance for the future than the fact that the last hailstorm occurred on a Tuesday, or is the writer indicating the existence of a structure, of an institution, that others could use as a guide to action in the future? When the intention appears to be the latter, it is not always clear why that which is described should be regarded as a manifestation of something regular rather than of something quite fortuitous.

The question also presents itself when one considers the tables of statistics that often accompany such studies, or when a public speaker on an economic subject seems to feel incumbent to read out reams of figures. Perhaps it is supposed that the readers or listeners, on letting a series of figures pass through their mind, will somehow acquire a more substantial understanding of the past than if they were given only an account in terms of what various people tried to do, or in terms of their greed or generosity, their envy, anger, compassion, loyalty, and so on; an account that most people, one suspects, would find a good deal more intelligible. Perhaps the writer or speaker, unable to find much order or regularity in the figures, presents them to the readers or listeners in the hope that they can see some significance in them. Perhaps, again, statistics are provided simply so that they may be stored away on a bookshelf or in the back of the mind as something that may or may not come in useful one day.[1]

In short, it is sometimes not at all clear whether the studies under discussion are meant to provide a genetic understanding of the subject matter, or to arrive at *ex ante* facts by induction. This apparent lack of methodological clarity is perhaps not very surprising since deterministic equilibrium models, based on the view that explanation and prediction are the inverse of each other, cannot be expected to give much theoretical guidance in the matter. It may be worthwhile to consider whether the empirical studies of descriptive economics would be better served by a theoretical framework which does allow choice to be analysed in an *ex post* context. While it will not be possible to do this here, I shall take a step in that direction by considering briefly the most obvious question to arise out of the conclusions stated in the first paragraph of this section, namely, what difference would it make to micro-economic theory if choice could appear in it either as *ex ante* or as *ex post* facts? In section 6.2 I shall consider a method of analysis that has considered choice in an *ex post* context and in section 6.3 I shall suggest that a generalized form of this method is not incompatible with what has at times been a major preoccupation of economic theory. Section 6.4 will deal with the question of consistent choice and economic institutions.

6.2 The method of *Verstehen* as a form of the genetic understanding

The genetic understanding, it will be recalled, depends on a set of beliefs. An explanation of the past, or an account of how something in the present here and now has come to be there, is intelligible to us if it conforms to our beliefs. When we interpret the behaviour of others, we most commonly do so in the belief that they act purposefully as we do and that they have emotions and sentiments similar to our own. It is thus an analogy to ourselves that mainly makes such interpretations meaningful to us and creates *ex post* facts in the form of surmised aims, expectations, sentiments, and so on. Furthermore, the aims, sentiments, and so on that we ascribe to others, being analogous to our own, provide us with an understanding that is intuitive, direct and intimate or, as is often put, with an understanding rather than with a mere explanation. Since most people, at least in a fairly sophisticated milieu, would ascribe aims and sentiments only to human beings (and possibly to some animals) this intuitive understanding is not available to us when we are dealing with purely physical processes.

This conclusion, namely, that human action but not nature may be understood intuitively or directly, has been used to draw a sharp epistemological distinction between the natural sciences on the one hand and history, the social and the 'cultural' sciences on the other. The former, it has been argued, have to rely on regularities found by induction whereas the latter can and do explain events as manifestations of the thought, intentions and sentiments of purposefully acting and feeling individuals, without necessarily any recourse to regularities found by induction. The historian or the social scientist can, so to say, re-enact a past event in his or her mind – making use what is often called the method of *Verstehen* (*Verstehen*, German for understanding, has in this context the connotation of empathy, of feeling and imagining oneself into the position of another).

A version of this viewpoint was put forward by Vico in the 18th century, but attracted little attention. The view gained acceptance, however, when it was put forward more than a century later as an analytical approach to the philosophy of history, that is, as an alternative to speculative philosophies of history (for example, that history follows some set pattern, or is the unfolding of providential wisdom, or of logical necessity as in Hegel and Marx). The term *Verstehen* is associated with Wilhelm Dilthey, but similar views were expressed by Benedetto Croce, whom I have mentioned before, the Oxford philosopher R.G. Collingwood and others. The distinction between generalizing and individualizing procedures, which I mentioned in connection with Menger, became in the hands of Rickert and other neo-Kantian philosophers very similar to Dilthey's, and emphasized that *Verstehen* was concerned with concrete, unique situations and with the values that individuals attached to them. The method of *Verstehen* also influenced the Austrian school in economics through von Mises and led to Max Weber's formulation of a '*verstehende*' sociology.

The method of *Verstehen* merely gave a respectable academic status to a very familiar procedure; and this had the advantage of encouraging an investigation of its implications. Previously the idea had been applied as a matter of common sense. Adam Smith, for instance, made 'changing places in fancy' a central principle of his *Theory of Moral Sentiments* (which was based on the same lecture notes from which the *Wealth of Nations* evolved). 'As we have no immediate experience of what other men feel', he pointed out, 'we can form no idea of the manner in which they are affected, but by conceiving what we ourselves should feel in the like situation'.[2] In the more formal development of the concept of *Verstehen*, the emphasis has tended to be on purpose and rationality. Smith integrated purpose fully with the emotive aspects of action and in this he seems merely to have been following common sense. One can certainly understand an action by reconstructing a cool calculation of what constituted adequate means to some immediate end, but in interpreting an action one usually goes beyond this to the sentiments attached to some more ultimate goal. Two observers may be agreed upon an immediate means-ends relation, yet they would understand the action quite differently if one sees it (say) as serving an idealistic struggle for liberty and the other as serving a ruthless grabbing for power. Understood broadly, the method of *Verstehen* simply consists of linking the visible effects of action with an imagined state of mind. It is of course practised by all and sundry in their daily dealings with other people and in their efforts to make sense of human affairs in general.[3]

When a study in descriptive economics involves the actions of, say, manufacturers, trade-unionists, government officials and politicians, it is thus natural that the investigator should changes places in fancy with the subjects being studied in order to gain an understanding in the sense of *Verstehen*. Indeed, the study may seem colourless and off-target if this is not done. Provided the investigator also follows other canons of good historiography, a study in descriptive economics is then really an historical monograph on recent economic events, in principle little different from the monographs produced by Schmoller's historical school.

Let us suppose that we regard such historical or ideographic (see Chapter 4, section 4.5) monographs as a very worthwhile form of economic analysis. We know from the inconclusiveness of the *Methodenstreit* that this attitude does not commit us to questioning the value of an abstract and more theoretical economics. However, mindful of the bitterness of the *Methodenstreit*, and of the attacks by the historical schools on classical political economy, and of various institutionalists on more recent theory, we suspect that the criticisms of theory could not have been entirely vacuous. We would therefore like to have a theory that is as much as possible in harmony with ideographic procedures, or better still, one that can be a theoretical framework for descriptive studies of the actual course of events, as is surely the ideal. One way of achieving this, it seems to me, is to generalize, or abstract from, the method of historical enquiry discussed in this section in such a way that one is left with an abstract form of the genetic understanding.

In the process of abstraction the fullness or richness in which concrete situations can be understood by empathy (or by the method of *Verstehen*) must necessarily be lost. The great diversity of particular motives and sentiments must necessarily give way to a more abstract concept. The concept of utility, of use value, or of the German *Nützlichkeit*, as originally used by economists, seems to me to be such a concept. Its scope was not thought to encompass all motives and sentiments, though something more than the desire for wealth in political economy. Exactly what it did encompass, which amounts to the question of how economic action is to be distinguished from other action, has always been surrounded by a lot of haziness. However, the attempt to remove the last 'vestigial traces of the utility concept' from economics (Chapter 5, section 5.7) would also remove the last traces of an intuitive *Verstehen* from economics. One would then be committed to dealing only with consistent choice, or, if one seeks comprehensiveness of explanation, with preference fields.

6.3 The abstract genetic understanding in economics

There are many versions of a generalized or abstract genetic understanding, and some of these rely on an intuitive *Verstehen*. Hegel's dialectical unfolding of the mind or spirit is usually held to be the most ambitious of these. Marx's philosophy of history, which applies the dialectic to the productive relations among men, is not much less ambitious, even if more down to earth. Both, of course, are what are normally called speculative systems. On the purely analytical level the contributions have been more modest. I have already tried to show that Menger intended his exact economics to be an abstract scheme for analysing historical processes. Max Weber's ideal types can be put to the same use. In short, there is hardly a lack of material for a study of the abstract genetic understanding. However, I shall not even attempt to draw up a list of the available material. I shall confine myself to indicating one possibility for adapting deterministic equilibrium models so that they may yield an abstract genetic understanding which, I shall argue, economists have at times intended their subject to yield.

For these purposes we require an abstract model that retains only the most general features, in non-specific form, of an explanation of how something found in the present has come to be there. In order to see what such a model may look like, it may be useful to look for some parallels. Let us take biological evolution by natural selection as an exemplar of a genetic explanation. We are interested only in its most general features and therefore the fact that it is an example from the natural rather than the social sciences will be seen, I think, not to matter.

Of the assumptions needed for explaining the existence of a particular species, the following are of interest in the context:

1. There are life-forms which reproduce themselves and which are vulnerable to their environment, that is, those ill-adapted to their environment die.
2. Offspring inherit genes which determine and limit their capacity for developing physical characteristics.
3. Gene mutations occur which lead to inheritable variations.
4. The physical, including the climatic, environment changes.

Now, the point that I want to bring out is that the theory of evolution is an explanatory model and not a deterministic model. It can explain how a species has evolved but it cannot predict what species will evolve in the future. The reason for this is plain. Items 3 and 4, mutations and environmental changes, are unpredictable. Within fairly broad limits anything could happen. On the other hand, in order to be an explanatory model the theory must contain some constraints on what could possibly have happened in the past. These constraints are listed as items 1 and 2. For example, characteristics acquired during the life of an organism could not have been transmitted to offspring. Equipped with such an explanatory model, the biologist can then undertake a genetic explanation which he or she can make as detailed as desired and for which the existence of fossil evidence will give certain fixed points. Since items 1 and 2 place certain constraints on the explanation, the biologists will be able to infer the changes that took place under items 3 and 4, and will be able to infer that at some stage in the past a certain inheritable variation must have appeared or that the environment must have changed in a certain way. In other words items 3 and 4 are mere explanatory 'shells', so to say, which acquire specific forms only during the course of the explanation. They are empty shells for forming *ex post* facts.

We may now use evolution by natural selection as an exemplar of a genetic explanation. On that basis we may conclude that an explanatory model in economics need not be a deterministic model and that it must consist of explanatory constraints (without which explanation is impossible) and explanatory 'shells' for forming *ex post* facts. We may also find some fixed points for our explanations in the form of documentary or other evidence from the past. The ideas of explanatory constraints and shells and of fixed points thus form the abstract model of a genetic explanation that we set out to find.

Let us lower the level of generality for a moment. The explanatory constraints may now take on the form of economic institutions, such as the institutions of a market economy or of (a particular stage of) feudalism. On the other hand, the belief that other people have aims, expectations and sentiments, just as we do, provides us with mere explanatory shells. We may believe that other people have aims, expectations, and so on, but we do not know what they are aiming at or what they expect, and so on. This can only be inferred during the course of a constrained explanation, that is, it is only in drawing up an intelligible account of the actions of others that their specific aims, expectations and sentiments are formulated as *ex post* facts. As is well-known, economic institutions are continually undergoing change and we are therefore at a disadvantage, as compared to the biologist, because our explanatory constraints are not as stable as the biologist's. Nevertheless we have to be able to formulate some constraints in order to explain the genesis of an event. Even though we can make use of the method of *Verstehen*, which the biologist cannot, the variety of aims, expectations and sentiments that it would be possible to infer from some event (that is, as having led to the event) would be almost unlimited if no constraint were placed on the explanation.

It may now be seen that an explanatory model in which choice appears as an *ex post* fact differs from a deterministic model in so far as the direction of explanation is inverted. Instead of starting with an initial or earlier state and determining a final or later state, we have to start with a final or later state and infer an initial or earlier state. The difference is more than just a curiosity. It means that the ways value and price are conceived in deterministic and explanatory models are subtly but significantly different. Let me try to illustrate this first with respect to the concept of efficiency. It is sometimes said that a free market mechanism eliminates inefficient firms. Now, in the context of a deterministic model this statement is tantamount to a prediction that efficient firms will survive. In the context of an explanatory model it means that firms which survive may be called efficient. In the deterministic sense, efficiency is a property which is already inherent in the firms in the initial state, and it will determine their fate in the ensuing course of events. With an explanatory model we have to infer the initial state from the final state, in this case the fact that certain firms have survived. The efficient firms are those that best adapted themselves to the cost and demand conditions which happened to prevail in the preceding period. There can be no other criterion for efficiency than that a firm survived. It may seem that this makes a tautology of efficiency; it is like saying that a race was won by the winner. It is, however, not a tautology. The criterion for efficiency in a market economy is in fact survival. Firm A may do all things according to books on management, but if it goes bankrupt while firm B, which skimps along on very low overheads, survives, it would be said that firm B was the more efficient under the conditions that prevailed. The argument for the market economy is in fact not that efficient firms survive, but that the *meaning and value judgement* attached to the word efficiency may be applied to surviving firms. In a different economic system it may possibly be said that the concerns that survive are the ones that are favoured by the Party, and we would understand (*Verstehen*) the matter quite differently. The method of *Verstehen* retains an influence in genetic explanations but loses it in deterministic models.

There is a similar difference between the way value and price are understood in a deterministic model and in a genetic explanation. In a deterministic model, prices are the result of various behavioural equations or *ex ante* facts, including the preference fields of individuals. In a genetic explanation, choice is an *ex post* fact and a tautology may seem to arise again, namely, that the consumer prefers whatever he or she happens to choose. But it is again not a tautology. The original argument was that under the conditions of a market economy with its free choice and free competition, the thing that an individual happens to choose can be understood as something valued, and the term maximum utility (signifying something like the most desirable state of affairs attainable) could be applied to the common outcome of all choices. *Verstehen* had become an abstract genetic understanding.

Walras explained the rationale of his general equilibrium analysis in terms which made this point. At the end of Chapter 5, section 5.4 I quoted Walras as saying that he had not attempted to predict decisions, but only to express the effects of such decisions. In the same Lesson he also dealt with the 'importance of a scientific formulation of pure economics'. He had treated free competition as an hypothesis and for this it was unimportant whether one actually 'observed it in the real world', as long as one could 'form a conception of it'. 'It was in this light that we studied the nature, causes and consequences of free competition. We now know that these consequences may be summed up as the attainment, within certain limits, of maximum utility. Hence free competition becomes a principle or a rule of practical significance.' He acknowledged that this was the argument for laissez-faire that economists had used for a long time. He differed from them, he thought, in that he had actually proved the argument. 'Nevertheless, I should like to ask: how could these economists prove that the results of free competition were beneficial and advantageous if they did not know just what these results were?' He illustrated the advantages of rigorous analysis with the by-now-familiar argument that the idea of maximum utility could not be applied to communally consumed goods nor to goods produced by natural monopolies. It also could say nothing about what is called distributive justice.[4]

Walras, as I have pointed out, usually opted for the *ex post* context when he had to make a decision. The concept of a preference field did not arise with him. The utility analysis was new, it was more rigorous, but the idea was an old one. When Adam Smith spoke of 'that general objection which may be made to all the different expedients of the mercantile system; the objection of forcing some part of the industry of the country into a channel less advantageous than that in which it would run of its own accord', one can well imagine his adding: 'Whatever that may be.'[5] If he had lived later he might have said 'Whatever it may be, it would represent maximum utility.'

6.4 Consistent choice and institutions

So far I have been mainly concerned with choice as an *ex post* fact, or with the explanatory shells in the alternative to a deterministic model. I now want to consider the explanatory constraints in that model. In the previous section I simply posited that these constraints take the form of economic institutions. An institutional set-up in which free competition takes place was then seen to affect the meaning of the *ex post* facts of choice.

How have institutions normally been conceived by economic theorists? Are they really accorded the status of constraints on action? The following quotation from Schumpeter seems to me to sum up the traditional attitude, or at least the attitude when the question is paid any attention at all.[6]

> The schemata of economic theory derive the institutional frameworks within which they are supposed to function from economic history, which alone can tell us what sort of society it was, or is, to which the theoretical schemata are to apply. [Furthermore] ... when we introduce the institution of private property or of free contracting or else a greater or smaller amount of government regulation, we are introducing social facts that are ... a sort of generalized or typified or stylized economic history.

There can be no doubt that historical studies can provide one with an appreciation of what are normally called institutions that is richer, more satisfying and more meaningful than any other comprehension one may have of them. To say so is merely to acknowledge the primacy of a genetic understanding, or more particularly of the method of *Verstehen*. Nevertheless, I would maintain that a genetic understanding of institutions is inappropriate to the needs of economic theory.

The institutional framework for economic theory appears to be an amorphous and ill-specified bundle of things, and a repository for anything not easily handled by theory. However, let us consider briefly what generalizations one may make about institutions. Schumpeter went on to suggest that one could define human behaviour widely enough to include 'the social institutions that are relevant to economic behaviour such as government, property inheritance, contract, and so on'. If one stretches the meaning of choice a little, it would seem that a great many of the forms of institutions are really cases of consistent choice, in the sense of ends and the choice of means, and constraints on choice which, in the limit of Hobson's choice, coincide with consistent choice. So, customs and conventions, regulations with legal sanction, norms of good behaviour and so on, can qualify for the appellation of institution. Consistent choice and constraint on choice do not cover all cases; numbers of people and geographical proximity, for instance, seem to have a role in market forms. However, it was my contention in Chapter 2 that anything with the lasting qualities necessary to qualify for the term institution may end up described by an unspecified number of *ex ante* facts. Consistent choice is merely a special case.

As we have seen, however, when the question of what institutions actually exist does arise among theorists, the information is to be obtained not by inductive methods but from economic history. Understood historically or genetically, institutions are *ex post* facts. If the theories for which they are to be the framework are deterministic models, then choices in general appear in them as *ex ante* facts, in the guise of consistent preference fields. Now, this scheme seems to me to be the inverse of what it should be if economics is to have an empirical content. As I have tried to show in Chapters 4 and 5, there appears to be no good reason for supposing that choices in general could ever be formulated as *ex ante* facts, but that is how they appear in deterministic models.

On the other hand, there appears to be at least a chance that institutions could be formulated as *ex ante* facts, but we are to understand them as *ex post* facts. Can any really worthwhile attempt be made to isolate institutional data empirically with such a conceptual scheme? Can one, for instance, look for consistent choice when all choice is meant to be taken care of already by the preference field idea?

Let us consider the famous law of demand. Here, it seems, is a very rare case of a constant conjunction in economics, even if it is somewhat erratic by the standards of the physical sciences. The preference field idea, at least in Hicks's view (Chapter 5, end of section 5.7), was developed in order to account for the law of demand, though other hypotheses would also have been possible. (It would therefore be naive to ask how one can deny the existence of preference fields and yet accept the law of demand, which is meant to arise out of certain common features of preference fields.) Let us suppose, for the sake of argument, that one were to put forward a rival life-style hypothesis (Chapter 5, end of section 5.6) to account for the same observed regularities. It would proceed on the principle suggested by Duesenberry, namely, that a 'real understanding of the problem of consumer behaviour must begin with a full recognition of the social character of consumption patterns'.[7] The details are here not important. The hypothesis may simply be based on a 'demonstration effect' or it may be a complex of the common ends, ultimate values, normative orientations, and so on, that arose out of Talcott Parsons's analysis.[8] As long as there were certain complementary and commonly held aims and more than one good could serve a particular aim, there would be enough complementary goods and substitutes to 'explain' the observed 'qualitative' relations between prices and quantities sold.

Anyone who put forward such a hypothesis would no doubt be faced by an army of objectors (if he or she was taken seriously enough). People do not all live up to the same life-style, incomes differ, most individuals have idiosyncrasies, and so on, in comparison, the preference field idea would be extremely neat and clean, especially if preference fields are allowed to vary from time to time. However, would it necessarily be a superior explanatory device? The life-style analysts would be talking about things that everyone knows something about and is therefore able to criticize. The preference field analysts are talking about a catch-all which no one has experienced, and it is in the nature of a catch-all that it can explain everything and nothing.

Notes

[1]   The usefulness of mere statistical records is of course equally limited in the natural sciences. In a newspaper interview (*The Star*, Johannesburg, 10 August 1976, p 50, from the Guardian News Service) an earth scientist used a homely analogy to make this point. Commenting on the question of when the next earthquake may be expected in San Francisco, he said: 'We can analyse records and use our equipment, but however sophisticated your equipment it can't predict the future. Your car milometer will tell you how many miles you've driven last year but it can't say how many you will drive next year.' Keynes, in the days before the General Theory, went further and warned against the tendency for statistical induction to be confused with mere statistical description, such as the fitting of a trend line by the method of least squares (*A Treatise on Probability* [London: Macmillan, 1929], pp 327–9). Von Mises never tired of pointing out that statistics are history.

[2]   Adam Smith, *The Theory of Moral Sentiments* (London: Bell, 1892), p 3.

[3]   The method of *Verstehen* is in fact so much part of our everyday lives that it often seems to escape attention altogether, or, possibly on the principle that familiarity breeds contempt, that there is a reluctance to accord it the status of a respectable intellectual method. It has been my good fortune to know Professor Lachmann who over the course of some years has brought home to me the central position of *Verstehen* in human affairs, as well as the tenor of Austrian economics. I owe him a great deal of debt for this and for much else that I could not enumerate here. Afternoon tea at the Lachmanns' has always been to me not only a most convivial occasion but also a truly academic experience. I should mention that Professor Lachmann has, I think, some misgivings about the feasibility of establishing what I have called *ex ante* facts in the social sciences, and more generally about my attempt to reconcile what appear to be the aspirations of the mainstream of neo-classical economists with the approach of the Austrians. Nevertheless, those who know Professor Lachmann will easily recognize his influence.

[4]   L. Walras, *Elements of Pure Economics*, translated from the edition of 1926 by William Jaffe (London: Allen & Unwin, 1954) pp 255–7.

5   Adam Smith, *An Inquiry into the Nature and Causes of the Wealth of Nations*, Cannan edn (New York: Random House, 1937 [1776]) p 482.

6   J.A. Schumpeter, *History of Economic Analysis* (London: Allen & Unwin, 1954), p 20.

7   J.S. Duesenberry, *Income, Saving and the Theory of Consumer Behavior* (Cambridge, MA: Harvard University Press, 1949) p 19. Duesenberry's study of dependence, and especially his discussion in the earlier parts of the book, really amount to a life-style hypothesis. However, his formulation of the utility function retains the comprehensiveness of deterministic models.

8   Talcott Parsons, *The Structure of Social Action* (Glencoe: McGraw-Hill, 1937). The concepts appear throughout his study. He draws up tentative conclusions on pp 727–53.

Select Bibliography

Arrow, K.J. and Hahn, F.H. (1971) *General Competitive Analysis.* Edinburgh: Holden-Day, Oliver & Boyd.

Ayres, C.E. (1951) 'The Co-ordinates of Institutionalism', Papers and Proceedings of the American Economic Association, *American Economic Review*, 41: 47–55.

Becker, G.S. (1962) 'Irrational Behaviour and Economic Theory', *Journal of Political Economy*, 70: 1–13.

Callie, W.B. (1964) *Philosophy and the Historical Understanding.* London: Chatto.

Cohen, M.R. and Nagel, E. (1934) *An Introduction to Logic and Scientific Methods.* London: George Routledge and Sons.

Devons, E. (1961) 'Applied Economics: The Application of What?' in *The Logic of Personal Knowledge, Essays Presented to Michael Polanyi on his Seventieth Birthday.* London: Routledge.

Duesenberry, J.S. (1949) *Income, Saving and the Theory of Consumer Behavior.* Cambridge, MA: Harvard University Press.

Eddington, A. (1939) *The Philosophy of Physical Science.* Cambridge: Cambridge University Press.

Ferguson, C.E. (1969) *Microeconomic Theory*, revised edn. Homewood: Richard D. Irwin.

Galbraith, J.K. (1967) *The New Industrial State.* London: Hamilton.

Hahn, F.H. (1973) *On the Notion of Equilibrium in Economics.* Cambridge: Cambridge University Press.

Hayek, F.A. (1949) *Individualism and Economic Order.* London: Routledge.

Hayek, F.A. (1949) 'The Use of Knowledge in Society' in *Individualism and Economic Order.* London: Routledge, pp 77–91.

Hayek, F.A. (1967) *Studies in Philosophy, Politics and Economics.* London: Routledge.

Hicks, J.R. (1956) *A Revision of Demand Theory.* Oxford: Clarendon.

Hume, D. (1739) *A Treatise of Human Nature.* London: John Noon.

Hutchison, T.W. (1977) ' "Crisis" in the Seventies: The Crisis of Abstraction' in *Knowledge and Ignorance in Economics.* Oxford: Blackwell, pp 62–97.

Jevons, W.S. (1957 [1871]) *The Theory of Political Economy*, 5th edn. New York: Kelley & Hillman.

Lachmann, L.M. (1976) 'From Mises to Shackle: An Essay on Austrian Economics and the Kaleidic Society', *Journal of Economic Literature*, 14: 55–62.

Laplace, P.-S. (1951) *A Philosophical Essay on Probabilities*, translated from the 6th French edn by F.W. Truscott and F.L. Emory. New York: Dover.

Marshall, A. (1920) *Principles of Economics*, 8th edn. London: Macmillan.

Menger, C. (1883) *Untersuchungen über die Methode der Socialwissenschaften*. Leipzig: Duncker. Reprinted by the London School of Economics as vol II of *The Collected Works of Carl Menger*, 1933. The book has been edited by L. Schneider and translated into English by F.J. Nock as *Problems of Economics and Sociology* (Champaign: University of Illinois Press, 1963).

Menger, C. (1933–6) *The Collected Works of Carl Menger*. London: London School of Economics Reprint.

Mill, J.S. (1872 [1843]) *A System of Logic*, 9th edn, vol II. London: Longmans.

Nagel, E. (1961) *The Structure of Science*. London: Routledge.

Pareto, V. (1906) *Manual of Political Economy*. First published 1906, translated from the French edn of 1927 by A. Schwier (London: Macmillan, 1972).

Pareto, V. (1935) *The Mind and Society*, translated from the 1923 edition by A. Bongiorno and A. Livingston. London: Cape.

Polanyi, M. (1958) *Personal Knowledge*. London: Routledge.

Popper, K.R. (1959) *The Logic of Scientific Discovery*. London: Hutchinson. Original German edition 1935.

Robbins, L. (1935) *An Essay on the Nature and Significance of Economic Science*, 2nd edn. London: Macmillan.

Rosenberg, N. (1960) 'Some Institutional Aspects of the Wealth of Nations', *Journal of Political Economy*, 68: 557–70.

Samuelson, P.A. (1973) *Economics*, 9th edn. New York: McGraw-Hill.

Schumpeter, J.A. (1943) *Capitalism, Socialism and Democracy*. London: Allen & Unwin.

Shackle, G.L.S. (1972) *Epistemics and Economics*. Cambridge: Cambridge University Press.

Shackle, G.L.S. (1974) 'Decision: The Human Predicament', *The Annals of the American Academy of Political and Social Science*, 412: 1–10.

Smith, A. (1892 [1759]) *Theory of Moral Sentiments*. London: Bell.

Smith, A. (1937 [1776]) *An Inquiry into the Nature and Causes of the Wealth of Nations*. Cannan edn. New York: Random House.

Tagliacozzo, G. (1945) 'Croce and the Nature of Economic Science', *Quarterly Journal of Economics*, 59: 307–32.

von Mises, L. (1931) 'Vom Weg der Subjektivistischen Wertlehre', in L. von Mises and A. Spiethoff (eds) *Probleme der Wertlehre*. Leipzig: Duncker, pp 73–94.

von Mises, L. (1949) *Human Action*. London: Hodge.

von Mises, L. (1958) *Theory and History*. London: Jonathan Cape.

von Mises, L. (1962) *The Ultimate Foundations of Economic Science*. Princeton: van Nostrand.

von Wieser, F. (1929) *Gesammelte Abhandlungen*, edited by F. von Hayek. Tübingen: Mohr.

Walras, L. (1954) *Elements of Pure Economics*, translated from the edition of 1926 by W. Jaffé. London: Allen & Unwin.

Index

References to endnotes show both the page
number and the note number (231n3).